Experimental Mechanics

29th Symposium on Experimental Mechanics in memory of Prof. Jacek Stupnicki, 19-22 October 2022, Warsaw, Poland

Editors
Paweł Pyrzanowski
Mateusz Papis

Peer review statement

All papers published in this volume of "Materials Research Proceedings" have been peer reviewed. The process of peer review was initiated and overseen by the above proceedings editors. All reviews were conducted by expert referees in accordance to Materials Research Forum LLC high standards.

Published under License by **Materials Research Forum LLC**
Millersville, PA 17551, USA

Published as part of the proceedings series
Materials Research Proceedings
Volume 30 (2023)

ISSN 2474-3941 (Print)
ISSN 2474-395X (Online)

ISBN 978-1-64490-258-1 (Print)
ISBN 978-1-64490-259-8 (eBook)

This book contains information obtained from authentic and highly regarded sources. Reasonable efforts have been made to publish reliable data and information, but the author and publisher cannot assume responsibility for the validity of all materials or the consequences of their use. The authors and publishers have attempted to trace the copyright holders of all material reproduced in this publication and apologize to copyright holders if permission to publish in this form has not been obtained. If any copyright material has not been acknowledged please write and let us know so we may rectify in any future reprint.

Distributed worldwide by
Materials Research Forum LLC
105 Springdale Lane
Millersville, PA 17551
USA
https://www.mrforum.com

Manufactured in the United State of America
10 9 8 7 6 5 4 3 2 1

Table of Contents

Preface

The 29th Symposium on Experimental Mechanics was held October 19-22, 2022 in Warsaw. It was organized on behalf of the Institute of Aeronautics and Applied Mechanics, Warsaw University of Technology; Committee on Mechanics of the Polish Academy of Sciences and Polish Association for Experimental Mechanics. The conference is organized every 2 years, since 2006 it is in memory of prof. Jacek Stupnicki - one of the most known polish scientists in the field of experimental mechanics. Due to the epidemiological situation, the Symposium was not held in 2020.

The main purpose of the Symposium is to enable researchers to present their latest experimental achievements in mechanics of solids, machine design, mechanical engineering, biomechanics. Starting from 2022, the Symposium's topics have also included issues related to fluid mechanics.

In 2022 the conference was attended by 58 participants including 17 students and PhD students. The participants came from Poland, Slovakia, the Czech Republic and France. Best papers were selected by the Scientific Committee for full-length publication.

Committees

HONORARY COMMITTEE

Marek Bijak-Żochowski
Lech Dietrich
Małgorzata Kujawińska
Józef Szala

SCIENTIFIC COMMITTEE

Paweł Pyrzanowski - Chairman
Romuald Będziński
Dariusz Boroński
Witold Elsner
Aniela Glinicka
Jerzy Kaleta
Maria Kotełko
Zbigniew Kowalewski
Grzegorz Milewski
Tadeusz Niezgoda
Piotr Paczos
Dariusz Rozumek
Magdalena Rucka
Leszek Sałbut
Jacek Szumbarski

ORGANIZING COMMITTEE

prof. Paweł Pyrzanowski – Chairman
Irena Mruk, Ph.D. – Honorary Secretary
Mateusz Papis, Ph.D.
Łukasz Klotz, Ph.D.
Przemysław Klik, M.Sc.
Dawid Maleszyk, M.Sc.

Experimental Mechanics
Materials Research Proceedings 30 (2023) 1-6

Materials Research Forum LLC
https://doi.org/10.21741/9781644902578-1

Optimization of parameters during filament extrusion

Tomáš Balint[1,a*], Jozef Živčák[1,b], Miroslav Kohan[1,c], Samuel Lancoš[1,d], Bibiána Ondrejová[1,e]

[1] Biomedical Engineering and Measurement Department, Faculty of Mechanical Engineering, Technical university of Košice, Letná 9, 042 00, SK

[a]tomas.balint@tuke.sk, [b]jozef.zivcak@tuke.sk, [c]miroslav.kohan@tuke.sk, [d]samuel.lancos@tuke.sk, [e]bibiana.ondrejova@tuke.sk

Keywords: Filament Extrusion, 3D Printing, PLA, 3D Printer, Filament Maker

Abstract. This scientific study brings new insights into the field of optimization of parameters in the extrusion of filaments from biodegradable materials. Extrusion is a production process in which metal or plastic materials are pushed through a rigid cross-sectional profile or matrix to form a continuous strip of shaped product (filament). The extrusion process begins by bringing the material in the form of granules, pellets or powders from the hopper to the extruder zone. One of the chapters contains a detailed description of the extrusion of filaments and optimization of parameters. Optimization of parameters consists of real designs and devices designed by the authors of this publication themselves. This study has a significant contribution in the field of material extrusion.

Introduction

Natural and synthetic polymers are considered biodegradable materials. Polymers can be broadly defined as macromolecules composed of covalently bonded monomers. Natural-based polymers include starch, chitosan, hyaluronic acid derivatives, collagen, fibrin gels, and silk. Undesirable properties of these polymers include low mechanical strength, unknown degradation rate, repellency and high physiological activity [1-4]. Various scientific studies show that synthetic polymers have a wide range of uses and satisfactory properties compared to natural polymers. An overview of biodegradable materials used in the extrusion of filaments and their possible applications is given in table 1.

Synthetic biodegradable polymers	Application
Poly(amino acids)	Medical products, tissue engineering, orthopaedic applications
Polymlinic acid (PLA), polyglycollic acid (PGA) and copolymers	Barrier membranes, controlled tissue regeneration (in dental applications), orthopaedic applications, stents, clamps, stitches, tissue engineering
Polyhydroxy butyrate (PHB), polyhydroxyvalerate (PHV) and copolymers	Long-term drug administration, orthopaedic applications, stents
Polydioxanone (PDO)	Fracture fixation, stitches

Table 1. Applications of biodegradable polymers

Extrusion is a manufacturing process in which metal or plastic materials are forced through a solid cross-sectional profile or die to form a continuous strip of shaped product (filament). The extrusion process begins with the introduction of material in the form of granules, pellets or powders from the hopper into the extruder zone. The melting process then begins through the heat generated by the mechanical energy supplied by the rotation of the screw and the heaters located

Experimental Mechanics | Materials Research Forum LLC
Materials Research Proceedings 30 (2023) 1-6 | https://doi.org/10.21741/9781644902578-1

along the head. The molten materials are then pressed into a die, which structures the materials into a hard pipe during the cooling process [5, 6].

Extrusion systems for the production of bioresorbable materials for clinical use

The extrusion manufacturing process is widely used for mixing polymeric materials. The process is highly flexible and enables a high degree of personalization of production. In twin-screw extrusion, the screws can be, for example, parallel or counter-current, interlocking or vice versa. In addition, the configurations of the augers themselves can be changed using various elements, blocks, to achieve specific mixing characteristics. In this extrusion process, raw materials can be solid substances (granules, powders). Large companies use industrial filament makers (extrusion systems), which excel in the large volume of extruded material and its precision. But in my research I will be using desktop extrusion systems at a good level. The desktop extrusion systems mentioned below are getting closer and closer in terms of quality to industrial extrusion systems. The Filabot EX2 extrusion system has a maximum extrusion temperature of 450°C, which means that all kinds of materials can be extruded, including high-temperature ones such as polycarbonate and even PEEK. Filament maker from 3devo company is an available extruder with a maximum extrusion temperature of 300°C and an extrusion speed of 250-600 mm per minute. We chose this extruder as the best option for extruding materials PLA/PHB/Thermoplastic starch and different concentrations of plasticizer, due to the precise composition of this extruder, the quality of the extruded materials and the extrusion temperature [7-10]. I am researching these materials as part of my dissertation. Extruder with high precision and maximum extrusion temperature of 300°C. Extrusion systems are mentioned in the following Fig. 1.

Fig. 1. Extrusion systems [10]

In industrial production, which uses more complex extrusion systems, unlike single-screw extrusion, a number of materials, both solid and liquid, are extruded by parallel two-screw extrusion (Fig.2). This provides maximum extrusion flexibility by allowing materials to be introduced into the melt at different stages or locations along the extruder cylinder. This brings a number of advantages:

• Possibility to add fibrous material to minimise fibre wear;

• Addition of shear or temperature-sensitive materials that can deteriorate if they pass through the entire extruder;

Experimental Mechanics

Materials Research Forum LLC

Materials Research Proceedings 30 (2023) 1-6

https://doi.org/10.21741/9781644902578-1

• Adding a plasticizer, liquid dye, stabiliser or lubricant.

Industrial extrusion systems are different from desktop extrusion systems in design and build quality. The barrel and screw of industrial extruders are made of high quality alloy steel with high hardness, strong corrosion resistance and long service life after nitrogen treatment. The automatic hydraulic inverter of the sieve can maintain the continuous production of the machine [11-14].

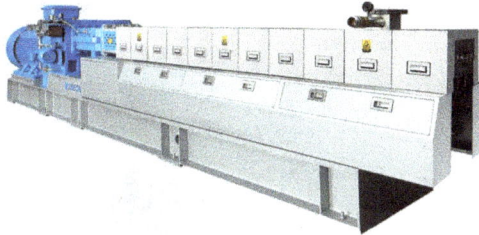

Fig. 2. Industrial extrusion system [12]

Twin-screw extruders are commonly used in modern industry. Based on the relative direction of rotation of their screws, these extruders can be divided into two types: co-rotating and counter-rotating. In a co-rotating twin-screw extruder, the maximum speed is reached at the tips of the screw, while in counter-rotating twin-screw extruders, the maximum speed is reached in the area of the feed. However, the counter-current mechanism generates a greater increase in pressure, making it more efficient at extrusion. Single- and twin-screw extruders were compared by a scientific team led by scientist Senanayake as part of design research aimed at a simplified extruder for less developed countries. The advantage of single-screw extruders is their simplicity of construction, but they are more likely to become clogged with material than twin-screw extruders. Furthermore, the single-screw extruder is the most common type of extruder and offers relatively low investment costs for companies dealing with the extrusion of materials intended for biodegradable purposes. If higher production and higher performance are required, twin-screw extruders are used. The easiest way to increase the throughput of the extruder is to increase the speed of the screw. This easy solution usually results in poor melt quality caused by exceeding the melting capacity of the screw design and degradation caused by high melt temperature. Using a smaller diameter screw can offer several advantages to achieve higher throughput at a higher screw speed [15].

Optimization of parameters during biomedical filament extrusion

After thorough research into materials and extrusion systems, we proceeded to extrude the biomedical filament on equipment from the 3devo company. The filament was produced from medically certified pellets in an optimized laboratory environment in a laminar box designed and engineered by this scientific team. Fig. 3 shows the filament production process.

Fig 3. Filament production process

We managed to optimize all necessary extrusion parameters. The laminar box equipped with a filter and cooling has fulfilled its purpose (Fig. 4).

Fig. 4. Biomedical laminar box

As part of improving the quality of the production of biomedical filaments, we proceeded to our own laminar box design. We made the model of the laminar box in the 3D modeling program SketchUp. The laminar itself stands out with its simple and purposeful design. Compact dimensions ensure trouble-free handling when operating the filament maker in the production process. A sterile environment is ensured by isolation. It is made of aluminum profiles and plexiglass with a thorough connection. Air recovery in the laminar box is provided by twelve fans that supply and remove air. Laminar boxes can be equipped with UV light, which has the task of sterilizing the working environment of the laminar box. The UV light unit is equipped with a timer and a sensor that prevents exposure to UV radiation when the door is lifted [16].

Conclusion

This scientific study brings important significant knowledge in the field of materials, production and optimization of parameters in the filament production process. For a better orientation in the given issue, the extrusion of biomedical filament on a desktop device from the company 3devo is described. Filament maker Composer 450 is a device for the production of filaments, on which it is possible to mix several materials. Production took place in an air-conditioned room at a temperature of 18°C. The material for the production of biomedical filaments was delivered in the form of granules, vacuum-sealed in an opaque package. Nevertheless, we dried the material in a dryer from 3devo. We set the drying temperature at 160°C for 180 minutes. We optimized this entire process with the help of a device designed by us. The laminar box can also be defined as a laboratory station intended for work in dust-free, sterile conditions. The use of laminar boxes is wide-ranging. Laminar boxes are mainly used in optical, laser, semiconductor and electronic technology applications. The design of the laminar box is designed to prevent contact with the external environment and thus ensure the protection of researchers as well as the researched material. The entire article is briefly divided from the materials to the output in the form of a new product.

Acknowledgement
This scientific study was created thanks to support under the Operational Program Integrated Infrastructure for the project "Center for Medical BioaddITive Research and Production (CEMBAM), code ITMS2014 +: 313011V358, co-financed by the European Regional Development Fund" and also thanks to support under the Op-erational Program Integrated Infrastructure for the project: Open Scientific Community for Mod-ern Interdisciplinary Research in Medicine (OPENMED), code ITMS2014 +: 313011V455, co-financed by the European Regional Development Fund and thanks to support under the Opera-tional Program Integrated infrastructure for the project: Center for Advanced Therapies of Chronic Inflammatory Diseases of the Musculoskeletal System (CPT ZOPA), code ITMS2014 +: 313011W410, co-financed by the European Regional Development Fund.

References
[1] Zeeratkar, M., D. De Tulio, M., Percoco, G., 2021. Fused Filament Fabrication (FFF) for Manufacturing of Microfluidic Micromixers, an Experimental Study on the Effect of Process Variables in Printed Microfluidic Micromixers. In: Micromachines. Vol. 12(8), pp. 858. https://doi.org/10.3390/mi12080858

[2] Hasegawa, K.; Matsumoto, M.; Hosokawa, K.; Maeda, M., 2016. Detection of methylated DNA on a power-free microfluidic chip with laminar flow-assisted dendritic amplification. In: Anal. Sci., 32, pp. 603-606. https://doi.org/10.2116/analsci.32.603

[3] Ou, J.; Moss, G.R.; Rothstein, J.P., 2007. Enhanced mixing in laminar flows using ultrahydrophobic surfaces. In: Phys. Rev. E, 76, 016304. https://doi.org/10.1103/PhysRevE.76.016304

[4] A. Aghaei Araei, I. Towhata., 2014. Impact and cyclic shaking on loose sand properties in laminar box using gap sensors . In: Soil Dyn Earthq Eng. Vol. 66 , pp. 401-414. https://doi.org/10.1016/j.soildyn.2014.08.004

[5] Ueng TS, Chen CH, Peng LH, Li WC., 2006. Large-scale shear box soil liquefaction testing on shaking table - preparation of large sand specimen and preliminary shaking table test. In: Proceedings of the National Center for Research on Earthquake Engineering; Taiwan.

[6] M. Khabbazian, V.N. Kaliakin, C.L. Meehan., 2010. Numerical study of the effect of geosynthetic encasement on the behaviour of granular columns. In: Geosynthetics International. Vol. 17. pp. 132-143. https://doi.org/10.1680/gein.2010.17.3.132

[7] Tagliavini G., Solari F., Montanari R., 2016. CFD simulation of a co-rotating twin-screw extruder: Validation of a rheological model for a starch-based dough for snack food. In: Proceedings of the International Food Operations and Processing Simulation Workshop, FoodOPS. https://doi.org/10.1515/ijfe-2017-0116

[8] Pearson J.R.A., Petrie C.J.S. The flow of a tubular film. Part 1. Formal mathematical representation. J. Fluid Mech. 1970;40:1-19. doi: 10.1017/S0022112070000010. https://doi.org/10.1017/S0022112070000010

Experimental Mechanics   Materials Research Forum LLC
Materials Research Proceedings 30 (2023) 1-6   https://doi.org/10.21741/9781644902578-1

[9] Pearson J.R.A., Petrie C.J.S. The flow of a tubular film. Part 2. Interpretation of the model and discussion of solutions. J. Fluid Mech. 1970;42:609-625. doi: 10.1017/S0022112070001507. https://doi.org/10.1017/S0022112070001507

[10] Vlachopoulos J., Sidiropoulos V., 2017. Reference Module. In: Materials Science and Materials Engineering. Elsevier.

[11] Wilczyński K., Lewandowski A., Wilczyński K.J., 2012. Experimental study for starve-fed single screw extrusion of thermoplastics. In: Polym. Eng. Sci. 2012;52:1258-1270. doi: 10.1002/pen.23076. https://doi.org/10.1002/pen.23076

[12] Gautam A., Choudhury G.S., 1999. Screw configuration effects on residence time distribution and mixing in twin-screw extruders during extrusion of rice flour. J. Food Process. Eng. 1999;22: pp. 263-285. https://doi.org/10.1111/j.1745-4530.1999.tb00485.x

[13] Kao S.V., Allison G.R. Residence time distribution in a twin screw extruder. Polym. Eng. Sci. 1984;24:645-651. doi: 10.1002/pen.760240906. https://doi.org/10.1002/pen.760240906

[14] Altomare R.E., Ghossi P., 1986. An Analysis of Residence Time Distribution Patterns in A Twin Screw Cooking Extruder. In: Biotechnol. Vol.2, pp. 157-163. https://doi.org/10.1002/btpr.5420020310

[15] Gonçalves N.D., Teixeira P., Ferrás L.L., Afonso A.M., Nóbrega J.M., Carneiro O.S., 1849. Design and optimization of an extrusion die for the production of wood-plastic composite profiles. In: Polym. Eng. Sci. Vol. 55, pp. 1855. https://doi.org/10.1002/pen.24024

[16] Bahaa S., Mohammad, A. Et al., 2021. Gaining a better understanding of the extrusion process in fused filament fabrication 3D printing: a review. In: The International Journal of Advanced Manufacturing Technology. Vol.114, pp. 1279-1291. https://doi.org/10.1007/s00170-021-06918-6

Experimental Mechanics
Materials Research Proceedings 30 (2023) 7-15

Materials Research Forum LLC
https://doi.org/10.21741/9781644902578-2

Analysis of large deformations of long flexible bars

Artur Ganczarski[1,a*], Tomasz Gawlik[1,b]

[1]Cracow University of Technology, Jana Pawła II 37, 31-864 Kraków, Poland

[a]artur.ganczarski@pk.edu.pl, [b]t.gawlik96@gmail.com

Keywords: Large Deformation, Verification of Bending Test by FEM and Theory, Carbon Fibre Composite

Abstract. This work presents a comparison the results of the real deformation of a four-segment fly rod used to the feeder method with the results obtained from the theory and the FEM. The experiment of bending comprises preparation of the measuring path, in which the real fly rod is loaded by a series of forces subsequently changing both magnitude and inclination. The FEM model of the fly rod is based on the beam element and the variation of the cross-section is subjected to stepping approximation. The theoretical model takes advantage of the classical elliptic integral formulation applied to describe full curvature problem of long flexible bars. Dominant errors between the experimental data and numerical results come from essential difficulties in accurate measurement of the wall thickness as well as uncertainty of fibre carbon configuration.

Introduction

Fly rods, independently of their destination, are designed as double-, triple- or four-segmental or alternatively as telescopic ones. Recently, the majority of fly rods is made of carbon fibres, whereas glass fibre fly rods represent rather lower quality goods. However, in case of many carbon fibre fly rods, where the tip segment does not play essential role in carrying of load but serves only for signalization that a fish swallows fish hook, this is so called fly rod with the vibrating tip segment, the tip segment made of glass fibre is used just for to assure sufficient stiffness.

Generally, fly rods of all kinds work in elastic range, where large bending is dominant state, whereas torsion or shear effects are negligible. Bending fly rods subjected to large deformations ought to exhibit high strength as well elasticity range. Moreover, the good quality fly rods should deform following special scheme: the deformation of tip segment resembles approximately parabola which extends towards hand grip segment according to an increasing load, and in case of advanced deformations may exhibit straightening effect.

Fly rods usually dedicated for feeder method of fishing are from 2.7m to 4.0m in length and serve for throw a masses from several grams to even 250g. The most frequent lengths of such rods are equal to 3.3m, 3.6m and 3.9m.

From the structure theory point of view, the most essential problem consists in description of fly rod deformation under loading. In case when the fly rod is treated as beam/rod element, an adequate description of deformation requires: consideration of nonlinear formula for finite curvature and simultaneously lack of prismatic shape of segments, as well as nature of loading, which may change both magnitude and direction. Associated problems known in literature of structural mechanics are as follows: finite displacements of beams – see [4], post-critical compression of column – see [5] and bending of beam of finite curvature subjected to an inclined force – see [1]. In all cases solutions are expressed by elliptic integrals and deal with prismatic beam element under concentrated load, which direction stays fixed in the space (force directed to a pole). Fundamental difficulty in adaptation aforementioned solutions to analysis of rod deformation consists in lack of prismatic shape of fly rod segments, which turn out to be conical. As consequence, engineer designing fly rod has at least two approaches to the problem: either to treat fly rod as beam/rod of step like cross-section – see section on nonlinear theory of bending, or

Experimental Mechanics   Materials Research Forum LLC
Materials Research Proceedings 30 (2023) 7-15   https://doi.org/10.21741/9781644902578-2

to take advantage of one of commercial Finite Element packages – see ANSYS Workbench model presented in further section.

Experimental Investigations

Test stand, shown in Fig. 1, comprises a stand supporting the fly rod, inclined to the ground with 60°, and a stand with grip to attach of a cable pulley, serving to thread a fishing line. White rope determines horizontal line necessary for setting up position of the cable pulley. Markers located subsequently at 2, 4, 6, 8 and 11m away form a hand grip of fly rod are fixed by use of a measuring tape. Both the stand of fly rod and the stand of cable pulley are made of an oak wood elements joined by steel L profiles. Loading is realized by series of normalized weights 100, 200, 500 and 1000g.

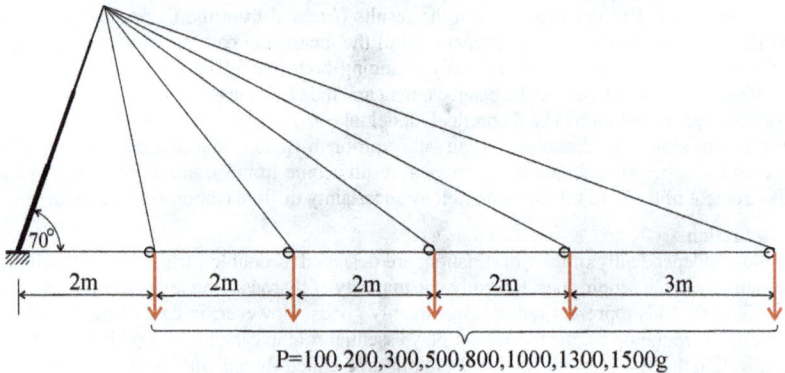

P=100,200,300,500,800,1000,1300,1500g

Fig. 1. Scheme of test stand

Experiment consists in registration by camera series of fly rod deformations referring to different combinations of load magnitude and distance measured with regard to the hand grip.

Initial test assumes following parameters: distance equal to 1.0 m, series of loading 100, 200, 300, 500, 800, 1000, 1300, 1500g and angle of inclination with respect to the ground equal 70°. This reflects final phase of towing when a fish is situated almost at the fisherman foot. Unfortunately, this test can be done for maximum weight 800g, since weight 1000g leads to failure of the fly rod. Aforementioned, negative result of initial test gives hints to the proper test characterized by following parameters: magnitude of angle of inclination with respect to the ground is decreased to 60°, minimal distance is increased to 2.0m.

Series of fly rod deformations under selected load magnitudes equal to 500, 1000 and 1500g referring to subsequent distances 11, 8, 6, 4 and 2 m are presented in Fig. 2.

Experimental Mechanics
Materials Research Proceedings 30 (2023) 7-15

Materials Research Forum LLC
https://doi.org/10.21741/9781644902578-2

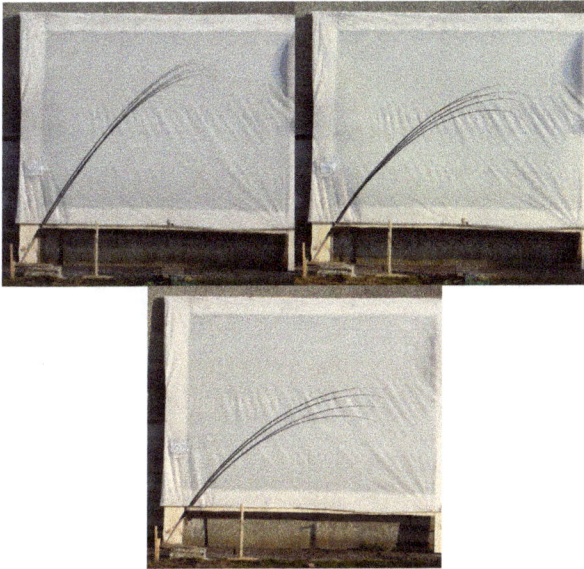

Fig. 2. Bending test of fly rod for selected load magnitudes equal to 500, 1000 and 1500g

Nonlinear Theory of Bending

Below the approach taken from monograph [1] for four-segment cantilever beam of step-like constant stiffness under concentrated force is presented – see Fig. 3.

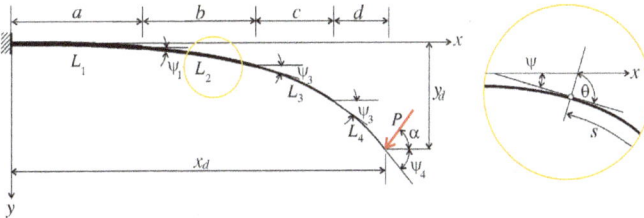

Fig. 3. Scheme of four-segment cantilever beam of step-like constant stiffness

Format of differential equation including magnitude of bending moment in current point (x,y) is following

$$EI_i \frac{d\psi}{ds} = M = P_1(x_d - x) + P_2(y_d - y), \tag{1}$$

where $P_1 = P \sin \alpha$ and $P_2 = P \cos \alpha$ are projections of force P according to subsequent axes, EI_i denotes bending stiffness of i-th segment, whereas s stands for coordinate measured along the arc – see window in Fig. 3. After differentiation with respect to s

$$EI_i \frac{d^2\psi}{ds^2} = -P_1 \cos \psi - P_2 \sin \psi, \tag{2}$$

and introducing new variables

$$u = s/L, \quad \theta = \psi + \alpha, \tag{3}$$

one can get

$$\frac{d\theta}{du} = L\frac{d\psi}{ds}, \quad \frac{d^2\theta}{du^2} = \frac{d}{du}\left(\frac{d\psi}{ds}\right)L = \frac{d}{ds}\left(\frac{d\psi}{ds}\right)L\frac{ds}{du} = \frac{d^2\psi}{ds^2}L^2, \quad \frac{d^2\psi}{ds^2} = \frac{1}{L^2}. \tag{4}$$

The right hand side of Eq. (2) can be rewritten as

$$P\left(\frac{P_1}{P}\cos\psi + \frac{P_2}{P}\sin\psi\right) = P(\sin\alpha\cos\psi + \cos\alpha\sin\psi) = P\sin(\psi+\alpha) = P\sin\theta, \tag{5}$$

whereas equation (2) itself takes format

$$\frac{d^2\theta}{du^2} + c_i\sin\theta = 0, \tag{6}$$

where $c_i = \frac{L^2 P}{EI_i} = L^2 k_i^2$. It is essential to emphasize here, that Eq. (6) is in fact the system of 4 nonlinear differential equations of second rank, which requires 8 conditions: 2 boundary conditions $+ 3 \times 2 = 6$ continuity conditions. Integration starts from 4-th tip segment using boundary condition

$$\psi(s=0)=0 \quad \text{and} \quad \left.\frac{d\psi}{ds}\right|_{s=0}=0 \quad \text{or} \quad \theta(u=0)=\alpha \quad \text{and} \quad \left.\frac{d\theta}{du}\right|_{\theta=\psi_4+\alpha}=0, \tag{7}$$

next multiplication of Eq. (6) for $i=4$ by $2d\theta$ and integration using boundary condition Eq. (7) yields

$$\frac{d\theta}{du} = \sqrt{2c_4[\cos\theta - \cos(\psi_4+\alpha)]}. \tag{8}$$

First of Eq. (3) yields

$$du = \frac{ds}{L} = \frac{d\theta}{\sqrt{2c_4[\cos\theta - \cos(\psi_4+\alpha)]}}, \tag{9}$$

hence integration with respect to ds from $\theta=\alpha$ to $\theta=\psi_3+\alpha$ gives

$$L_4 = \int_\alpha^{\psi_3+\alpha} ds = \frac{L}{\sqrt{2c_4}}\int_\alpha^{\psi_3+\alpha}\frac{d\theta}{\sqrt{\cos\theta - \cos(\psi_4+\alpha)}}, \tag{10}$$

or in equivalent format

$$\frac{L_4}{L} = \frac{1}{\sqrt{2c_4}}\left[-\int_0^\alpha\frac{d\theta}{\sqrt{\cos\theta - \cos(\psi_4+\alpha)}} + \int_0^{\psi_3+\alpha}\frac{d\theta}{\sqrt{\cos\theta - \cos(\psi_4+\alpha)}}\right]. \tag{11}$$

Since $\cos x = 1 - 2\sin^2\frac{x}{2}$, therefore Eq. (11) can be written down as sum of two elliptical integrals of first kind

$$L_4 = \frac{1}{k_4}[F(p_4) - F(p_4, m_4)], \tag{12}$$

where $m_4 = \sin^{-1}\left(\frac{1}{p_4}\sin\frac{\alpha}{2}\right)$ and $\sqrt{c_4} = Lk_4$.

In analogous manner, solutions for segment 3, 2 and 1 get format

$$L_i = \frac{1}{k_i}[F(p_i,\zeta_i) - F(p_i, \zeta_{i-1})], \qquad (13)$$

where $\sin\zeta_{i-1} = \frac{1}{p_i\sqrt{2}}$ and $\sin\zeta_i = \frac{1}{p_i}\sqrt{\frac{1+\sin(\psi_i+\alpha)}{2}}$.

It is convenient to compare number of equations and unknowns. There are 4 nonlinear equations (6) involving 5 unknowns ($\psi_4, p_4, p_3, p_2, p_1$), hence numerical procedure has to be preceded by trial/error estimation of ψ_4. When magnitudes of ($\psi_4, p_4, p_3, p_2, p_1$) are known next components of displacement vector can be calculated for each segment. Below solution for segment 4 (tip segment) is only demonstrated.

The infinitesimal length of arc according to Eq. (9) takes format

$$ds = \frac{d\phi}{k_4\sqrt{1-p_4^2\sin^2\phi}}, \qquad (15)$$

and after integration we get solution being the product of trigonometric functions and elliptical integrals of first and second kind

$$x = \frac{\cos\alpha[F(p_4,m_4)-F(p_4,n_4)+2E(p_4,n_4)-2E(p_4,m_4)]+2p_4\sin\alpha(\cos m_4-\cos n_4)}{k_4}, \qquad (16)$$

where $n_4 = \sin^{-1}\left(\frac{1}{p_4}\sin\frac{\psi+\alpha}{2}\right)$.

In similar way

$$dy = ds\cos\psi = ds\sin(\theta-\alpha) = \frac{\cos\alpha}{k_4}2p_4\sin\phi\, d\phi - \frac{\sin\alpha}{k_4}\left(\frac{d\phi}{\sqrt{1-p_4^2\sin^2\phi}} - \frac{2p_4^2\sin^2\phi\, d\phi}{\sqrt{1-p_4^2\sin^2\phi}}\right), \qquad (17)$$

finally leading to

$$y = \frac{2p_4\cos\alpha(\cos m_4-\cos n_4)-\sin\alpha[F(p_4,m_4)-F(p_4,n_4)+2E(p_4,n_4)-2E(p_4,m_4)]}{k_4}. \qquad (18)$$

To end this section authors invoke basic description of elliptical integrals taken from monograph [2]. Elliptical integral of first kind in Legendre's format is the function of variable ϕ and parameter p as follows

$$F(p,\phi) = \int_0^\phi \frac{d\vartheta}{\sqrt{1-p^2\sin^2\vartheta}} = \int_0^{\sin\phi}\frac{dt}{\sqrt{(1-t^2)(1-p^2t^2)}} \qquad (19)$$

whereas elliptical integral of second kind in Legendre's format is the function

$$E(p,\phi) = \int_0^\phi\sqrt{1-p^2\sin^2\vartheta}\, d\vartheta = \int_0^{\sin\phi}\sqrt{\frac{1-p^2t^2}{1-t^2}}\, dt \qquad (20)$$

where parameter p is called modulus of elliptical integral. Functions $F(p, \phi)$ and $E(p, \phi)$ are presented in tables and for real arguments p and $\sin \phi$ are subject to change in range between 0 and 1. Complete elliptical integrals of first or second kind are functions $K(p)$ or $E(p)$ of modulus p and variable $\phi = \pi/2$

$$K(p) = F(p, \pi/2), \; E(p) = E(p, \pi/2) . \tag{21}$$

Elliptical integrals of first and second kind have closed format only for $p=0$ and $p=1$, whereas in all other cases their values are calculated by expanding in appropriate series – see [3].

FEM Verification

The FEM model of the fly rod is based on the beam element and the variation of the cross section is subjected to stepping approximation – Fig. 4.

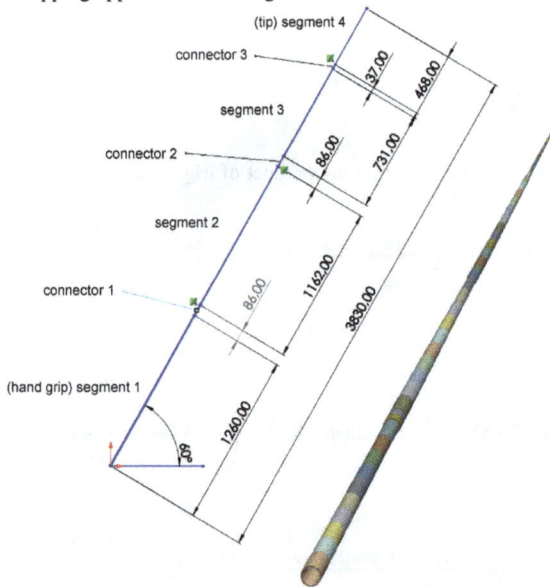

Fig. 4. Length of subsequent fly rod elements and its FE discretization

This approximation is based on division of each segment into ten equal elements and each connector, in which these segments link one to another, into two elements – Fig. 4.

Basic dimensions of the fly rod along perpendicular direction are small when compare to the length, hence for clearness they are collected in Tab. 1, where subsequent symbols stand for: l – length of segment, b – thickness, D_{max}/d_{max} – outer/inner diameter the biggest element in section, D_{min}/d_{min} – outer/inner diameter the smallest element in section, and additionally D/d – jump in diameter between two adjacent segments.

Experimental Mechanics
Materials Research Proceedings 30 (2023) 7-15

Materials Research Forum LLC
https://doi.org/10.21741/9781644902578-2

Tab. 1. Basic dimensions of fly rod elements in [mm]

element	number of elements	l	b	D_{max}	d_{max}	D_{min}	d_{min}	D/d
segment 1	10	126	1.2	19.2	16.8	14.24	11.84	0.63
segment 2	10	116	0.9	15.7	13.9	8.44	6.64	0.81
segment 3	10	73.1	0.75	9.4	7.9	5.22	3.9	0.46
segment4	10	46.8	–	3.5	–	1.6	–	0.21
connector1	2	43	2.1	16.04	11.84	15.7	11.5	0.34
connector2	2	43	1.65	9.94	6.64	9.4	6.1	0.54
connector3	2	18.5	–	5.22	–	5	–	0.22

Finite Element package ANSYS Workbench requires definition of material model, therefore the material Epoxy Carbon UD (230 GPA) is chosen from ANSYS library as the representative of carbon fibre/epoxy resin composite. Additionally, in order to improve material properties of aforementioned composite, the data presented in Tab. 2 is taken from the website of the producer of carbon fibre composite IM7 – see [7].

Tab. 2. Mechanical properties of carbon fibre composite IM7 [7]

typical hexplay 8552 composite properties (at room temperature)	US units	SI units	test method
0°tensile strength	395 [ksi]	2.723 [MPa]	
0°tensile modulus	23.8 [Msi]	164 [GPa]	ASTM D3039
0° tensile strain	1.6 [%]	1.6 [%]	
0° flexural strength	270 [ksi]	1.862 [MPa]	ASTM D790
0° flexural modulus	22.0 [Msi]	152 [GPa]	
0° short beam shear strength	18.5 [ksi]	128 [MPa]	ASTM D2344
0° compressive strength	245 [ksi]	1.689 [MPa]	ASTM Mod.
0° compressive modulus	21.7 [Msi]	150 [GPa]	D695
0° open hole tensile strength	62.1 [ksi]	428 [MPa]	ASTM D5766
0° open hole compressive strength	48.9 [ksi]	337 [MPa]	ASTM D6484
90° tensile strength	9.3 [ksi]	64.1 [MPa]	ASTM D3039
fibre volume	60 [%]	60 [%]	

Boundary conditions applied to FE model are as follows: (hand grip) segment 1 is fully clamped at the node referring to the origin of coordinate system, whereas (tip) segment 4 load by concentrated force at the final node.

Beam elements, shown in Fig. 4 (each colour refers to separate element), are conventional iso-parametric elements – see [6], whereas mesh size is set up as default. Numerical tests with manual remeshing confirm good convergence of FE code. In order to proper capture of large deformations the Automatic Load Displacement Control (ALDC) procedure is switched on.

Numerical simulations by FEM for selected loads 500, 1000 and 1500g and variable magnitude of distance are shown in Fig. 5.

Comparison of selected experimental results and corresponding FEM and theoretical results (see Fig. 6) is done by use of commercial software tool Kinovea dedicated to image analysis. Briefly speaking Kinovea creates system of reference lines attached to photos (experiment) as well as to figures (FEM), and as a consequence it allows user for correlation of results. In general, the attained correlation is good provided that if deformation is moderate, which means referring to small magnitudes of load (100–500g) and simultaneously long distances (11–6m).

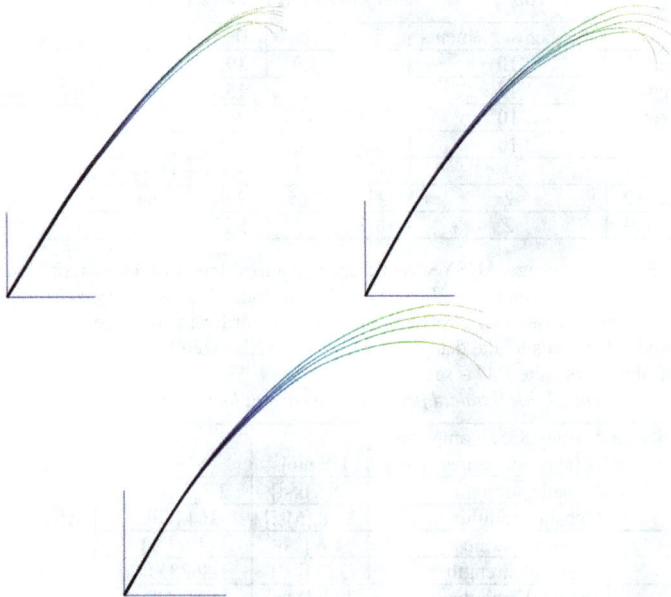

Fig. 5. FEM results for load magnitudes 500, 1000 and 1500g

On the contrary, in case of advanced deformations, referring to bigger magnitudes of load (800–1000g) and short distances (4–2m) some discrepancies are noticeable. Namely, central segments (#2 and #3) exhibit the biggest discrepancies for weight 1500g and distance 4m, whereas the tip element (#4) is responsible for generation of main discrepancies for weight 1000g and distance 2m.

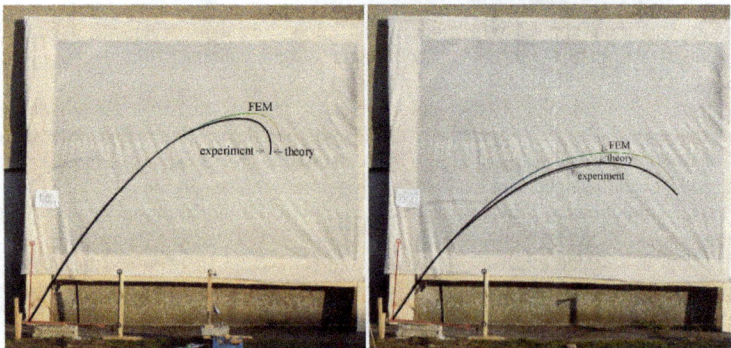

Fig. 6. Comparison of experimental, FEM and theoretical results for two selected combinations of load magnitudes and distances

Experimental Mechanics   Materials Research Forum LLC
Materials Research Proceedings 30 (2023) 7-15   https://doi.org/10.21741/9781644902578-2

In the opinion of authors of present work, there are two main sources of discrepancies:
- lack of information about the wall thickness, particularly with regard to central segments, that may essentially influence their stiffness,
- lack of honest information concerning carbon fibre configuration in the composite thus authors assumes orientation $0°$–$90°$ as a default.

Conclusions

Presented experimental data is well mapped by numerical results in case of moderate deformations, whereas major discrepancies observed for advanced deformations come from:
- essential difficulties in accurate measurement of the wall thickness,
- uncertainty of fibre carbon configuration, that is subject to commercial confidentiality.

Additionally, in case when even ALDC procedure (built in FEM package) fails the nonlinear theory of bending, taking advantage of elliptical integrals, is recommended for use.

References

[1] R. Frisch-Fay, Flexible bars, London Buttherworths, 1962.

[2] F. Oberhettinger, W. Magnus, Anwendung der Elliptischen Funktionen in Physik und Technik, Springer-Verlag, Berlin-Heidelberg, 1949. https://doi.org/10.1007/978-3-642-52793-7

[3] W. Press, S. Teukolsky, W. Vetterling, B. Flanner, Numerical Recipes in Fortran 77: The Art of Scientific Computing, Cambridge University Press, NY, 1997.

[4] S.P. Timoshenko, J.M. Gere, Mechanics of materials, Van Nostrand, New York, 1972.

[5] M. Życzkowski, Pokrytyczne zachowanie się prętów ściskanych, in: Mechanika techniczna, vol. IX, PWN, Warszawa, 1988, pp. 298–304.

[6] Information on http://riad.usk.pk.edu.pl/~m1/mysql/materialydydaktyczne/pliki/lkpizimes1.pdf

[7] Information on https://www.hexcel.com/user_area/content_media/raw/IM7_HexTow_Data Sheet.pdf

Experimental Mechanics

Materials Research Forum LLC

Materials Research Proceedings 30 (2023) 16-23

https://doi.org/10.21741/9781644902578-3

Parameter settings of the PEEK and PPSU filaments production with the ceramic component

Miroslav Kohan[1,a*], Samuel Lancoš[1,b], Tomáš Balint[1,c], Marek Schnitzer[1,d], Radovan Hudák[1,e] and Jozef Živčák[1,f]

[1] Department of Biomedical Engineering and Measurement, Faculty of Mechanical Engineering, Technical University of Košice, Letná 1/9, 042 00 Košice, Slovakia

[a]miroslav.kohan@tuke.sk, [b]samuel.lancos@tuke.sk, [c]tomas.balint@tuke.sk, [d]marek.schnitzer@tuke.sk, [e]radovan.hudak@tuke.sk, [f]jozef.zivcak@tuke.sk

Keywords: Filament, Extrusion, PEEK, PPSU, Hydroxyapatite, Tricalcium Phosphate

Abstract. The aim of the work is to determine suitable settings of filament production parameters for medical applications from materials PEEK and PPSU alone, as well as with a 10wt% of the ceramic components Hydroxyapatite (HA) and Tricalcium Phosphate (TCP) admixture. Filaments were made using the Filament Maker machine (3devo, The Netherlands). The filaments were manufactured according to the requirements for usage in the FDM technology with a nominal diameter of 1.75 mm. The diameter of the filaments was measured with an optical sensor and analyzed using DevoVision software (3devo, The Netherlands). The analysis of the filament diameters was carried out using descriptive statistics in order to determine quality of the filaments. The analysis of the produced filament diameter from the materials PEEK, PEEK + HA/ TCP , PPSU and PPSU + HA/ TCP demonstrated that the measured values of the diameters of the filaments from the nominal value (1.75 mm) showed minimal deviations, as well as the fact that the limit values were not exceeded (1.85 mm; 1.65 mm) and thus it is possible to state that the manufactured filaments meet the required quality for use in FDM technology. A microscopic analysis was also carried out on the manufactured filaments in order to determine the distribution of the ceramic component in the manufactured filaments. An Olympus GX71 inverted metallographic microscope with an Olympus DP12 camera was used for the purpose of expertise. The total number of examined samples was n = 40, while 10 samples from each filament were selected from random areas. Microscopic analysis of the produced filaments showed a uniform distribution of the ceramic component in the composite filaments, which means that the manufacturing process does not affect the distribution of the ceramic component in the filament.

Introduction

Fused Deposition Modeling (FDM) technology is part of additive manufacturing technology, which is gaining more and more use in various areas such as e.g. automotive production, cosmonautics or medical applications [1,2,3]. Statistical data indicate that the use of this technology in the field of medicine was at the level of 0.973%, but forecasts indicate that by year 2026 this share of use in the given sector will be at the level of 18.2% [4]. This fact creates new questions and requirements for the material side in the form of the production of new and high-quality filaments.

High demands are placed on the group of materials intended for medical applications in the form of biocompatibility. Polyetheretherketone (PEEK) and polyphenylsulfone (PPSU) can also be included in this group of biomaterials. PEEK is characterized by resistance to hydrolysis, high temperatures, wear and has good mechanical properties [5,6]. PPSU is characterized by high glass transition temperatures, high mechanical strength and stiffness, good chemical, hydrolytic and dimensional stability [7,8]. Based on these properties, PEEK and PPSU materials turn out to be suitable candidates for the replacement of biological structures in the form of bone. Another very

Experimental Mechanics   Materials Research Forum LLC
Materials Research Proceedings 30 (2023) 16-23   https://doi.org/10.21741/9781644902578-3

important parameter is osseointegration between the biomaterial and the biological structure in the form of bone. Due to the fact that PEEK itself does not have a biological activity for the process of osteonitration, it is necessary to create composite materials using Hydroxyapatite (HA) [9]. One such study is by the authors Senatov et al. [10] where they investigated the osseointegration level of scaffolds made of PEEK material with an admixture of HA in a cranial defect in mice. The results of the study show that scaffolds made of PEEK material with an admixture of HA demonstrated a higher level of osseointegration than pure PEEK scaffolds. A similar study by Durham et al. [11] investigated HA coating on PEEK implants in a rabbit model. Animals were studied in two groups of 9 for observation 6 or 18 weeks after surgery. The results of the study demonstrated that heat-treated HA coatings showed improved implant fixation as well as higher bone regeneration and bone-implant contact area compared to uncoated PEEK.

The aim of the subject study was to produce filaments with a diameter of 1.75 mm from the materials of pure PEEK and PPSU as well as variants with an admixture of a ceramic component in the form of HA and TCP that will meet the required production standards. The relevant analyzes in the form of the analysis of the diameter of the filament as well as the distribution of HA and TCP in composite materials represent the basic parameters of the quality of the manufactured filament. The output is to produce filaments that can be used in the 3D printing process and subsequently in pre-clinical studies.

Materials and methodology

Material characteristics

Single-component medical materials PEEK and PPSU as well as composite medical materials PEEK + HA/TCP and PPSU + HA/TCP were used for the production of filaments. All materials were in pellet form (see Fig. 1). In the composite material PEEK + HA/TCP, the mass ratio of the individual components was 80% PEEK, HA 10% and TCP 10%. In the composite material PPSU + HA/TCP, the mass ratio of the individual components was PPSU 80%, HA 10% and TCP 10%.

Fig. 1 Pellet form of materials for extrusion (A: pure PPSU ; B: pure PEEK ; C: PPSU + HA/TCP ; D: PEEK + HA/TCP)

Production of filaments

Filaments were produced from the materials described above on a Filament Maker Composer 450 (3devo, The Netherlands), which contains 4 heating zones. The entire production process consisted of 4 stages. In the initial stage, the Filament Maker was heated and cleaned at temperatures from 180 to 300 °C using HDPE and Devoclean Purge Mid cleaning materials. After cleaning, the filament production process continued with the 2nd stage where the filament production parameters were set (see Table 1). In the third stage of filament production, the diameter of the filament was recorded, which was determined to be 1.75 mm. The filament diameter was recorded via an optical sensor with an accuracy of ± 43 μm. After producing a sufficient amount of individual filaments, the process continued with stage 4 where the device was cooled and cleaned. The cooling of the device was up to a temperature of 180 °C with Devoclean Purge Mid and HDPE cleaning materials.

Table. 1 Basic settings for the production of PEEK filaments; PPSU; PEEK + HA/TCP; PPSU + HA/TCP

PEEK		PEEK + HA/ TCP	
Filament diameter	1,75 mm	Filament diameter	1,75 mm
Heat zone 1	390 °C	Heat zone 1	380 °C
Heat zone 2	390 °C	Heat zone 2	380 °C
Heat zone 3	385 °C	Heat zone 3	380 °C
Heat zone 4	375 °C	Heat zone 4	390 °C
RPM	4,5	RPM	3
Fan percentage	100 %	Fan percentage	100 %
PPSU		**PPSU + HA/ TCP**	
Filament diameter	1,75 mm	Filament diameter	1,75 mm
Heat zone 1	344 °C	Heat zone 1	344 °C
Heat zone 2	347 °C	Heat zone 2	347 °C
Heat zone 3	347 °C	Heat zone 3	347 °C
Heat zone 4	347 °C	Heat zone 4	347 °C
RPM	5	RPM	5
Fan percentage	72 %	Fan percentage	73 %

Microscopic analysis

Microscopic analysis was carried out using light and electron microscopy in order to quantitatively describe the distribution of the ceramic component. The total number of investigated samples was n = 40, while 10 samples were created from random areas from each filament produced. Before the analysis, the samples were prepared in dentacryl, sanded with sandpaper with a grain size of 200, 400, 600 and 800 μm. For the purposes of expertise, the experimental technique of light microscopy was used on the Olympus GX71 metallographic microscope with the Olympus DP12 camera. The details of the microstructure were checked by scanning electron microscopy on a Jeol JSM 7000F device in the mode of secondary electrons - SEI, which obtained detailed information about the morphology of ceramic particles and their distribution in the polymer filament matrix. Backscattered electron imaging - BSE provided information on the distribution of elements in the sample by atomic number. Areas in which elements with a higher atomic number are present are lighter in this display, on the other hand, areas formed by elements with a lower atomic number are shown as darker. A necessary condition for the analyzes of polymer matrix pellet samples was to ensure an electrically conductive surface of the preparation with each polymer matrix pellet. Before observation in the electron microscope, a layer of gold was deposited on all analyzed samples.

Statistical evaluation

The analysis of the diameter of the filaments was evaluated using descriptive statistics, while the following parameters were evaluated: diameter (x), standard deviation (SD), max./min. value, range (R), variance (Var(x)), kurtosis (K) and skewness (S_{KP}). Parameter K is an indicator of the distribution of measured data in the file. If its value is greater than 0, then the distribution is more peaked, and thus most recorded filament diameters approach the arithmetic mean.

Results

During the production of filaments from the given materials, the limit values of the diameter of the filament were set at 1.85 mm (upper limit) and 1.65 mm (lower limit). As a reference value for the diameter of the filament, a value of 1.75 mm was set. In Fig. 2 it is possible to see the graphs of the measured diameters of the produced filaments, where it is shown that these limit values of the diameters were not exceeded during the production process of the filaments. The average value for the produced filaments from the PEEK material was at the level of 1.747 ± 0.039 mm, for the PEEK + HA/TCP material at a value of 1.7473 ± 0.038 mm, for the PPSU material at 1.749 ± 0.039 mm and for the PPSU + HA/TCP material at a value of 1.749 ± 0.032 mm.

The SKP parameter evaluates the manufactured filaments from the point of view of the uniform distribution of the measured diameters of the filaments ($S_{KP} = 0$). If $S_{KP} > 0$, then smaller values prevail in the statistical set of measured filament diameters and the filament is thinner than the nominal value (1.75 mm). Conversely, if $S_{KP} < 0$, then higher values prevail in the statistical set of measured filament diameters and the filament is thicker than the nominal value. This indicator for the filament made of PEEK material had a value of $S_{KP} = 0.11$, which means that the filament is somewhat thinner than the nominal value. However, this minor deviation is not significant and does not affect the 3D printing process. The opposite effect was observed with other filaments. The S_{KP} parameter showed negative values (PEEK + HA/TCP = -0.04 ; PPSU = -0.06 ; PPSU + HA/TCP = -0.04) which indicates that more higher values than the calculated arithmetic mean were detected in the statistical set of measured averages. It is possible to state that with these filaments there are places with a larger diameter than the calculated arithmetic mean in the individual statistical files. However, these values represent minimal deviations from a uniform distribution, which can be considered insignificant. Other parameters can be seen in Fig. 2.

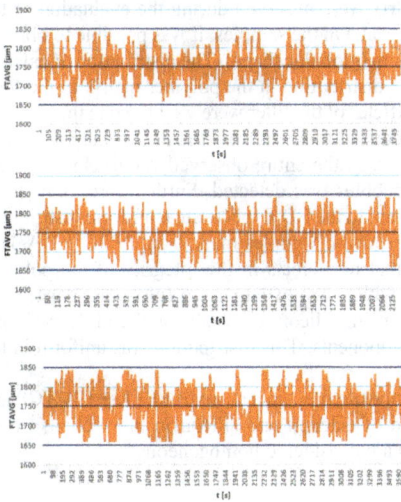

PEEK	
x = 1,747	R = 0,179
SD = 0,039	Var(x) = 1143,2
Max. = 1,84	K = 0,01
Min. = 1,661	S_{KP} = 0,11

PEEK + HA/TCP	
x = 1,7473	R = 0,179
SD = 0,038	Var(x) = 1504,1
Max. = 1,84	K = -0,43
Min. = 1,661	S_{KP} = -0,04

PPSU	
x = 1,749	R = 0,179
SD = 0,039	Var(x) = 1526,3
Max. = 1,84	K = -0,45
Min. = 1,661	S_{KP} = -0,06

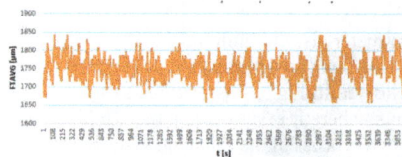

PPSU + HA/TCP	
x = 1,749	R = 0,179
SD = 0,032	Var(x) = 1049,93
Max. = 1,84	K = 0,12
Min. = 1,661	S_{KP} = -0,04

Fig. 2 Analysis of filament diameter of produced filaments using descriptive statistics (A: PEEK ; B: PEEK + HA/TCP ; C: PPSU ; D: PPSU + HA/TCP

Microscopic analysis

Fig. 3 shows the outputs in the form of light and electron microscopy for manufactured filaments from the materials PEEK, PEEK + HA/TCP, PPSU as well as PPSU + HA/TCP. During the evaluation of samples from PEEK filament, defects of the polymer matrix were not detected, and no filler was found in the matrix. In the EDX spectrum, the distribution of oxygen and carbon was visible in the entire section of the examined samples.

No defects in the integrity of the polymer matrix were observed during the evaluation of the produced filament made of PEEK + HA/TCP material. On the samples, it was possible to observe parts of the HA and TCP filler with a size of 1 to 2 μm (globular shape). The distribution of calcium and silicon was visible in the EDX spectrum. This distribution in the polymer matrix in the form of filler and clusters dispersed evenly, while all particles of the filler were well fixed in the polymer matrix. When evaluating the samples from the produced filament from the PPSU material, it was ensured that there is no filler in the polymer matrix in the entire observed detail. However, the presence of microscopic cracks in the form of bubbles was detected. Furthermore, only small scratches after sanding were observed in the close-up. In the EDX spectrum, a uniform distribution of oxygen and carbon was visible throughout the section. Samples made from PPSU + HA/TCP filament demonstrated by microscopic analysis that no defects in the integrity of the polymer matrix were detected. Irregular parts of the filler with a size of 1 to 2 μm (globular shape) were observed in the samples. The distribution of calcium and silicon was visible in the EDX spectrum, which in this case represents the HA and TCP components. This distribution was uniform in the section of the experimental sample.

Overall, it can be concluded that in the case of composite filaments made of PEEK + HA/TCP and PPSU + HA/TCP, there is a uniform distribution of calcium and silicon in the sections of the experimental samples, and therefore the filaments are considered homogeneous.

Materials Research Forum LLC

https://doi.org/10.21741/9781644902578-3

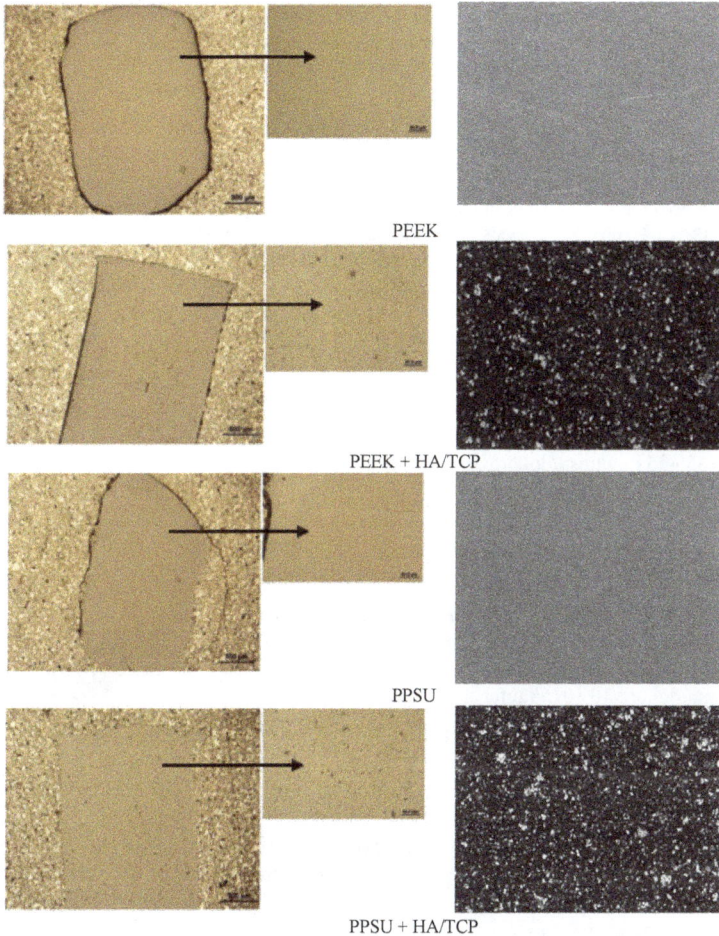

Fig. 3 Microscopic analysis of PEEK, PEEK + HA/TCP, PPSU and PPSU + HA/TCP materials

Acknowledgment
The achieved results were created within the investigation of the project no. 2018/14432: 1-26C0, which is supported by the Ministry of Education, Science, Research and Sport of the Slovak Republic within the provided incentives for research and development from the state budget in accordance with Act No. 185/2009 Coll. on incentives for research and development. This publication is the result of the project implementation CEMBAM - Center for Medical Bioadditive Research and Production, ITMS2014+: 313011V358 supported by the Operational Programme Integrated Infrastructure funded by the European Regional Development Fund. This publication is the result of the project implementation Open scientific community for modern interdisciplinary research in medicine (Acronym: OPENMED), ITMS2014+: 313011V455 supported by the Operational Programme Integrated Infrastructure funded by the European Regional Development Fund. This publication is the result of the project implementation Research and development of

intelligent traumatological external fixation systems manufactured by digitalization methods and additive manufacturing technology (Acronym: SMARTfix), ITMS2014+: 313011BWQ1 supported by the Operational Programme Integrated Infrastructure funded by the European Regional Development Fund.

Summary

Filaments made from the materials PEEK, PPSU, PEEK + HA/TCP and PPSU + HA/TCP meet the given regulations in the form of a diameter of 1.75 mm, as well as the homogeneity of the distribution of individual components in the case of composite materials in the form of PEEK + HA/TCP and PPSU + HA/TCP. Analysis of the diameter of the subject filaments showed minimal differences from the nominal value of the filament diameter of 1.75 mm. Microscopic analysis showed minimal clusters in the composite materials, which are considered insignificant and therefore it can be concluded that the produced filaments are homogeneous at the selected production parameters.

References

[1] SATHIES, T, Senthil P., Anoop M.S. A review on advancements in applications of fused deposition modelling process. Rapid Prototyping Journal [online]. 2020, 26(4), 669-687 [cit. 2022-11-23]. ISSN 1355-2546. https://doi.org/10.1108/RPJ-08-2018-0199

[2] SARTIPI, Farid, Kiran PALASKAR, Arman ERGIN, Uditha RAJAKARUNA. Viable construction technology for habitation on Mars: Fused Deposition Modelling. Journal of Construction Materials [online]. 2020, 1(2) [cit. 2022-11-23]. ISSN 26523752. https://doi.org/10.36756/JCM.v1.2.2

[3] DAMINABO, S.C., S. GOEL, S.A. GRAMMATIKOS, H.Y. NEZHAD, V.K. THAKUR. Fused deposition modeling-based additive manufacturing (3D printing): techniques for polymer material systems. Materials Today Chemistry [online]. 2020, 16 [cit. 2022-11-23]. ISSN 24685194. https://doi.org/10.1016/j.mtchem.2020.100248

[4] WICKRAMASINGHE, Sachini, Truong DO, Phuong TRAN. FDM-Based 3D Printing of Polymer and Associated Composite: A Review on Mechanical Properties, Defects and Treatments. Polymers [online]. 2020, 12(7) [cit. 2022-11-23]. ISSN 2073-4360. https://doi.org/10.3390/polym12071529

[5] PANAYOTOV, I. V. et al.: Polyetheretherketone (PEEK) for medical applications. In: Journal of Materials Science: Materials in Medicine. 27, 118, 2016. https://doi.org/10.1007/s10856-016-5731-4

[6] HALEEM, A. - JAVAID, M.: Polyether ether ketone (PEEK) and its 3D printed implants applications in medical field: An overview. In: Clinical Epidemiology and Global Health. 7, 4, 2019, s. 571-577. ISSN 2213-3984. https://www.sciencedirect.com /science /article/pii/S2213398418303178

[7] VICENTE BORILLE, A. - OLIVIERA GOMES, J. - LOPES, D.: Geometrical analysis and tensile behaviour of parts manufactured with flame retardant polymers by additive manufacturing. In: Rapid Prototyping Journal. 23, 1, 2017, s. 169-180. https://www.emerald.com/insight/content/doi/10.1108/RPJ-09-2015-0130/full /html

[8] SHUKLA, A. K. - ALAM, J. - ALHOSHAN, M.: Recent Advancements in Polyphenylsulfone Membrane Modification Methods for Separation Applications. In: Membranes. 12, 2, 2022, 247 s. DOI: 10.3390/membranes12020247. https://doi.org/10.3390/ membranes12 020247

Experimental Mechanics
Materials Research Proceedings 30 (2023) 16-23

Materials Research Forum LLC
https://doi.org/10.21741/9781644902578-3

[9] MA, Hongyun, Angxiu SUONAN, Jingyuan ZHOU, et al. PEEK (Polyether-ether-ketone) and its composite materials in orthopedic implantation. Arabian Journal of Chemistry [online]. 2021, 14(3) [cit. 2022-12-01]. ISSN 18785352. https://doi.org/10.1016/j.arabjc.2020.102977

[10] SENATOV, F., A. MAKSIMKIN, A. CHUBRIK, et al. Osseointegration evaluation of UHMWPE and PEEK-based scaffolds with BMP-2 using model of critical-size cranial defect in mice and push-out test. Journal of the Mechanical Behavior of Biomedical Materials [online]. 2021, 119 [cit. 2022-12-01]. ISSN 17516161. https://doi.org/10.1016/j.jmbbm.2021.104477

[11] DURHAM, John W., Sergio A. MONTELONGO, Joo L. ONG, Teja GUDA, Matthew J. ALLEN a Afsaneh RABIEI. Hydroxyapatite coating on PEEK implants: Biomechanical and histological study in a rabbit model. Materials Science and Engineering: C [online]. 2016, 68, 723-731 [cit. 2022-12-01]. ISSN 09284931. https://doi.org/10.1016/j.msec.2016.06.049

Experimental Mechanics
Materials Research Proceedings 30 (2023) 24-30

Materials Research Forum LLC
https://doi.org/10.21741/9781644902578-4

Influence of the purlin shape on the load-bearing capacity of sandwich panels

Monika Chuda-Kowalska

Poznan University of Technology, Poland, 60-965 Poznan, Pl. Sklodowskiej-Curie 5

monika.chuda-kowalska@put.poznan.pl

Keywords: Sandwich Panels, Purlin, Load-Bearing Capacity, Bending Test

Abstract. This study aimed to develop knowledge about the behavior of bent, multi-span sandwich panels. The analyzed panels have a soft polyisocyanurate foam core and rigid metal facings. The paper presents the results of experimental studies. The influence of support width, the span of the panel and purlin shape on the load-bearing capacity of the panel are analyzed. The tests carried out by the author have shown that not always the load capacity determined according to the standard is on the safe side. Therefore, the actual support conditions of the designed structure should always be taken into account.

Introduction

In this paper, sandwich panels composed of thin metal sheets and a thick, polyisocyanurate foam core (PIR) are considered. These kinds of structures are widely used in various areas of engineering. As a core material the different kinds of foams, usually made from polymers, metals, ceramics, glass, etc. are widely used in various branches of civil engineering since the 80s. In the literature, it is possible to find many papers focused on sandwich structures, their applications, designing [1-3] and testing procedures which take into account the influence of the soft core on the behavior of the layered structure [4, 5]. Nevertheless, because of the variety of factors affecting the structural response, e.g. variety of the core material, shape of the metal sheets, geometry and others, the development of design and testing methods is still a current challenge undertaken by scientists.

When analyzing this type of sandwich panels, despite the rather complex foam structure [6], it is usually assumed that the core material is homogeneous and isotropic or sometimes orthotropic. It is possible to use effective material parameters, which quite well reflect the global behavior of the panels [7]. Difficulties arise when local effects play a significant role, such as in the case of wrinkling of metal facing, where the stiffness of the core directly adjacent to the facing is decisive [8].

In this work, the main attention is focused on the analysis of the influence of the purlin shape and the deformation of the core material under the purlin on the load-bearing capacity of sandwich panels. All conclusions are drawn based on experimental tests.

Problem formulation

This paper deals with the problem of the behavior of bent, multi-span sandwich panels. At the intermediate support, an interaction of facing compression, core shear and compression of the core as a result of interaction with the support is observed.

In accordance with the EN 14509 standard [9], which is used in the design of sandwich panels, the experimental determination of the load capacity of multi-span panels above the intermediate support is simplified to a single-span scheme loaded with a linear load in the middle of the span (Fig.1). Such a static scheme, though simple, is very common in practice. It is more important, however, that the phenomena observed in a simple static scheme are easier to carry out, assess and interpret. In fact, in case of analyzed sandwich panels with soft core, the load transfer is more

Experimental Mechanics

Materials Research Proceedings 30 (2023) 24-30

Materials Research Forum LLC

https://doi.org/10.21741/9781644902578-4

complicated due to the variety of the material of the core, the thickness of the panel and the shape of purlin sections used. Simplifications of calculation schemes or assumptions are often used, however, they must give results on the safe side - the obtained limit loads should be close or lower than those in the actual structure.

The aim of the paper is to determine the influence of the core deformation on the load-bearing capacity of a sandwich panel depending on the length of the panel and the shape of the purlin. For this purpose, a 3-line bending test of the panel is performed, in which the load will be transferred by a steel beam (according to standard) in one case, and by the real shape of the purlin in the second case (the cold-formed thin-walled Z section).

a)

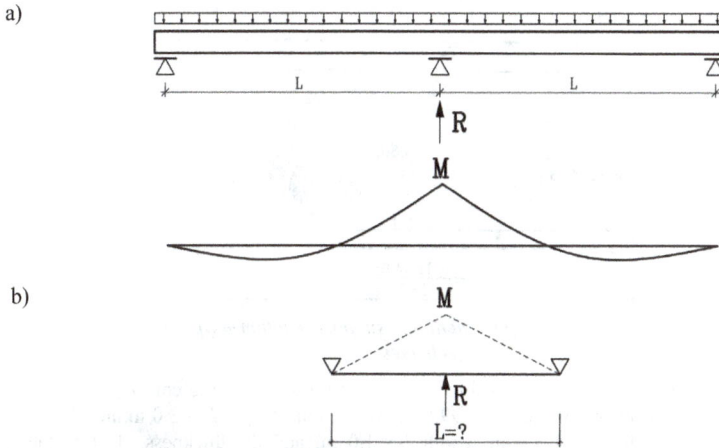

b)

Fig.1 a) actual static scheme of a multi-span beam, b) simplified static scheme - simulation of the intermediate support

Experimental approach

Sandwich panels with soft core are very sensitive to concentrated loads. Their deformation is more complicated and various phenomena such a bending, compression, delamination and shear is occurred what is shown in Fig.2.

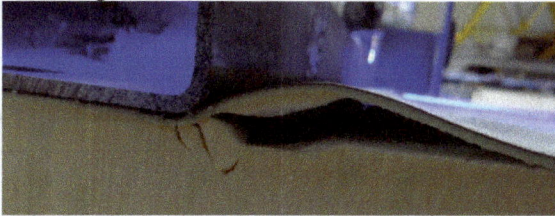

Fig.2 Forms of panel failure under concentrated load

The intensity of individual local failure mechanisms and their impact on the global behavior is closely related to the span of the panel, its bending stiffness (thickness of the core and their material parameters) and the way the load is applied. In order to accurately identify the behavior of the panel and the deformation of the foam core under the applied load, a series of full-scale tests are carried out.

Experimental Mechanics

Materials Research Proceedings 30 (2023) 24-30

Materials Research Forum LLC

https://doi.org/10.21741/9781644902578-4

In this work, sandwich panels with a PIR foam core with a density of 38 kg/m³ are tested. The first group of tests is planned with the aim to check the influence of the panel's span L_0 and the width of the applied load L_S on the local deformation of the core and load-bearing capacity of the whole panel. Therefore, three types of tests were carried out and analyzed, as shown in Fig.3. Schemes 1 and 2 will allow us to observe the effect of the width of the applied load. Schemes 1, 2 and 3 will be used to analyze of the impact of the span of the tested panel on its load-bearing capacity.

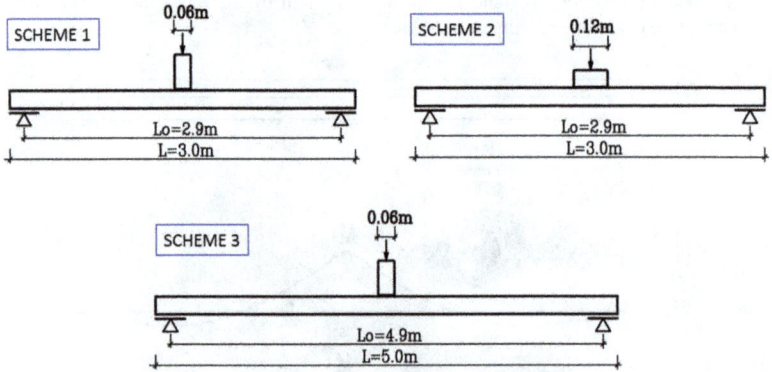

Fig.3 Interaction between bending moment and support reaction – types of experimental schemes

In this study, samples had following dimensions: the thickness of the core d_C = 99.24 mm, distance between the centroid of faces e = 99.67 mm, the total length L = 3.0 m and 5.0 m, the length of span L_0 = 2.9 m and 4.9 m, width B = 1.0 m, and the thickness of steel facings t = 0.43 mm. In each case, during the experiment, the applied force F and displacement u in the middle of the panel's span were continuously measured. Obtained results from all tests are summarized in Table 1.

Table 1: Experimental results of wrinkling stresses for Scheme 1 - 3

		Scheme 1	Scheme 2	Scheme 3
L_0	[m]	2.90	2.90	4.90
F_{max}	[N]	6352.60	7265.70	4243.30
$M(F_{max})$	[kNm]	4.61	5.27	5.20
σ_w	[MPa]	107.56	123.02	121.40

For maximum force F_{max} theoretical values of the wrinkling stress σ_w are calculated according to equation (1)

$$\sigma_w = \frac{M}{e \cdot t \cdot B} \, ,$$

where: $M = \frac{F_{max} \cdot L_0}{4}$.

(1)

Obtained results show, that for shorter plates, the width of the applied load plays a significant role. In the case of Scheme 1, the failure of the panel occurred surprisingly early, resulting in a very low value of the wrinkling stresses. If we use a wider support, changing the width L_S from

Experimental Mechanics
Materials Research Proceedings 30 (2023) 24-30

Materials Research Forum LLC
https://doi.org/10.21741/9781644902578-4

0.06 m to 0.12 m as in scheme 2, then the load-bearing capacity of the panel increases significantly (15 %) and is close to that obtained in Scheme 3.

For a more detailed analysis, the paths between bending moment and displacement measured in the middle of the span for each of the tests are presented in Fig.4.

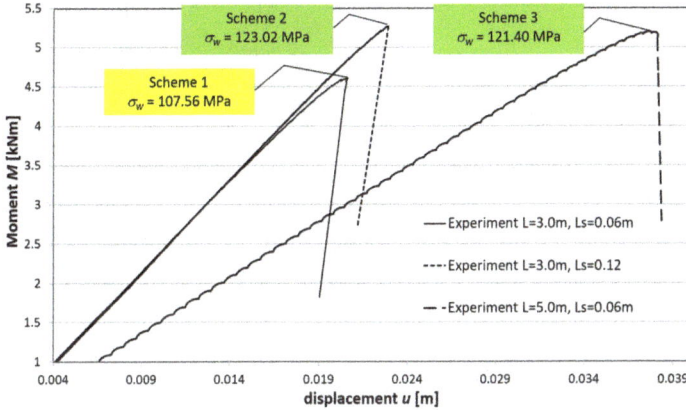

Fig. 4 Experimental paths obtained from three-point bending test

For higher load levels the plot for Scheme 1 reveals the non-linear response of the structure. The non-linear behavior is not manifested in the case of Scheme 2 or 3 because of lower stresses in the foam core directly under the applied load, consequently we observe smaller deformations of the core.

The second group of tests concerned the analysis of the influence of the purlin shape on the load-bearing capacity of sandwich panels. Therefore, two types of tests were carried out and analyzed, as shown in Fig.5. In this case, it was ensured that the width of the applied load was the same in both cases and equal to the minimum required standard value: $L_S = 0.06$ m.

In this study, samples had following dimensions: the thickness of the core $d_C = 99.01$ mm, distance between the centroid of faces $e = 99.47$ mm, the total length $L = 6.2$ m, the length of span $L_0 = 6.0$ m, width $B = 1.15$ m, and the thickness of steel facings $t = 0.46$ mm. In both cases, during the experiment, the applied force F and displacement u in the middle of the panel's span were continuously measured. Obtained results from both tests are summarized in Table 2.

Table 2: Experimental results of wrinkling stresses for Scheme 4 and 5

	Scheme 4	Scheme 5
L_0 [m]	6.00	6.00
F_{max} [N]	5480	4725
M [kNm]	8.22	7.09
σ_w [MPa]	156.22	134.69

Experimental Mechanics
Materials Research Proceedings 30 (2023) 24-30

Materials Research Forum LLC
https://doi.org/10.21741/9781644902578-4

a) Scheme 4

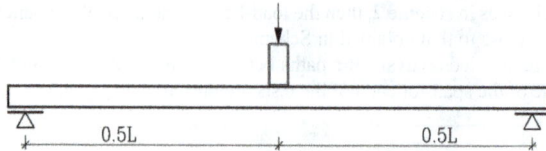

0.5L   0.5L

b) Scheme 5

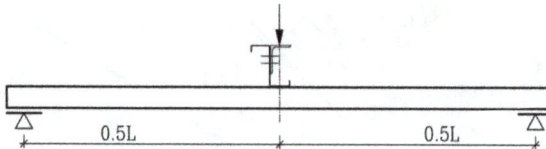

0.5L   0.5L

Fig.5 Purlin shape with $L_S = 0.06$ m: a) the box-section, b) the cold-formed Z-section

To analyze the obtained results we can say that the failure of both panels were occurred directly in the vicinity of the applied load, as shown in Fig.6. Additionally, the force-displacement paths, shown in Fig. 7, show a similar, linear behavior of both samples during the whole test until the failure. However, the load-bearing capacity of the panel loaded by box-section purlin is much higher (15 %) than in case of cold-formed Z section purlin. During the experiment, a slight rotation of the Z-section purlin was observed. Most likely, this caused uneven pressure of the purlin flange on the panel and thus accelerated the global damage to the whole sandwich panel.

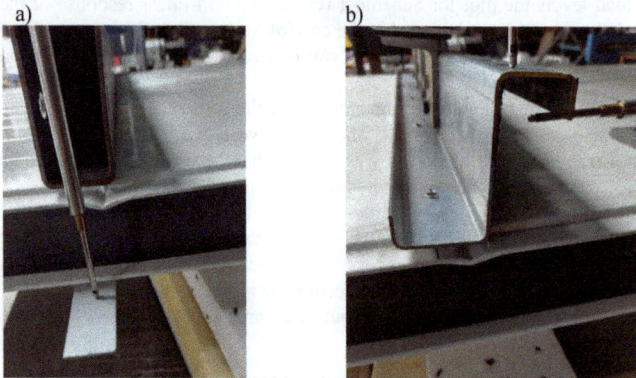

Fig. 6 Local deformation under the purlin: a) box section, b) cold formed Z section

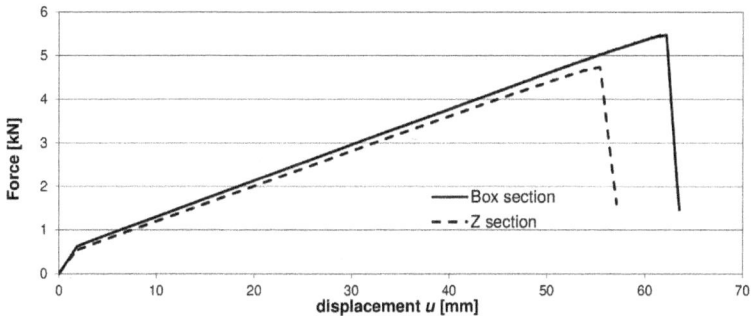

Fig. 7 Experimental paths F-u obtained for the box-section and the cold formed Z-section of the purlin

Summary
Different effects have an influence on the behavior and load-bearing capacity of sandwich panels with soft core. In this work, the main attention has been focused on the study of the structural sensitivity to the influence of the local deformation of the foam core, as well as the influence of the purlin shape. Sandwich plates loaded by concentrated loads exhibits a complex behavior. Their deformation are more complicated and various phenomena such a bending, delamination and shear occur. The obtained results clearly show that the higher stresses under the loading beam lead to faster failure of the panel. The load-bearing capacity of panels with a small span is strictly dependent on the width of the applied load. In this case, local deformations of the core can significantly reduce the load capacity of the panel. The load-bearing capacity of the panel on the central support given in the manufacturer's tables is determined for rigid purlins such as a box section. The tests carried out by the author have shown that not always the load capacity determined according to the standard [9] is on the safe side. The actual support conditions of the designed structure should always be taken into account.

Acknowledgements
The research was financially supported by Poznan University of Technology 0411/SBAD/0001

ORCID iD
Monika Chuda-Kowalska http://orcid.org/0000-0002-7250-6348

References

[1] J.M. Davies, (Editor), Lightweight Sandwich Constructions, Blackwell Science Ltd., 2001. https://doi.org/10.1002/9780470690253

[2] D. Zenkert, An Introduction to Sandwich Construction, EAMS, 1995.

[3] Z. K. Awad, Optimization of a sandwich beam design: analytical and numerical solutions, Structural Engineering and Mechanics, 48(1), (2013), 93-102. https://doi.org/10.12989/sem.2013.48.1.093

[4] R. Gibson, A simplified analysis of deflections in shear deformable composite sandwich beams, Journal of Sandwich Structures and Materials, 13(5), (2011), 579-588. https://doi.org/10.1177/1099636211408254

[5] S. Long, X. Yao, H. Wang, X. Zhang, Failure analysis and modeling of foam sandwich laminates under impact loading, Composite Structures, 197, (2018), 10-20. https://doi.org/10.1016/j.compstruct.2018.05.041

Experimental Mechanics
Materials Research Proceedings 30 (2023) 24-30

Materials Research Forum LLC
https://doi.org/10.21741/9781644902578-4

[6] M. Chuda-Kowalska, A. Garstecki, Experimental study of anisotropic behavior of PU foam used in sandwich panels, Steel and Composite Structures, 20(1), (2016), 43-56. https://doi.org/10.12989/scs.2016.20.1.043

[7] M. Chuda-Kowalska, Effect of foam's heterogeneity on the behavior of sandwich structures, CEER, 4(29), (2019), 097-111. https://doi.org/10.2478/ceer-2019-0047

[8] J. Pozorska, Z. Pozorski, Analysis of the failure mechanism of the sandwich panel at the supports, Procedia Engineering 177, (2017), 168-174. https://doi.org/10.1016/j.proeng.2017.02.213

[9] EN 14509, Self-supporting double skin metal faced insulating panels – Factory made products – Specifications, 2013.

Experimental Mechanics

Materials Research Proceedings 30 (2023) 31-38

Materials Research Forum LLC

https://doi.org/10.21741/9781644902578-5

Increasing the efficiency and flexibility of laboratory testing with virtual instrument techniques

Roland Pawliczek [1,a] *

[1]Faculty of Mechanical Engineering, Opole University of Technology,
Mikolajczyka 5, 45-271 Opole, Poland

[a] r.pawliczek@po.edu.pl

Keywords: Measurements, Control System, Virtual Instrument, Graphical Programming Environment

Abstract. This paper presents an upgrade of the functionality and modernization of the laboratory testing process using virtual instruments. A case study of airflow laboratory stand for air velocity profile determination and the fatigue testing on the MZGS100 stand shows the applications, where standard sensors and transducers are used as measuring devices. The article focuses mainly on DAQ (Data Acquisition) measurement techniques, where at present the USB communication method is very widely used. The main advantage of the system is the so-called open user interface, which is software developed according to the researcher's own algorithms. The developed software is just this virtual instrument, and the graphical programming environment is used as an effective tool to build the program. Virtual instrumentation based laboratory equipment present cost-effective, compact, and user-friendly human-machine interfaces for the measurement and laboratory equipment control.

Virtual instrumentation

It is a well-known fact that any scientific activity that requires experimental validation is associated with measurements and measuring devices. Measuring physical quantities requires systems equipped with sensors that convert these parameters into current quantities, and appropriate transducers that process the measured signals and present measurement results. Existing computer techniques also make it possible to transmit data to workstations and process them further using additional tools, software etc. This basic structure of measuring systems is called a "traditional" measuring device (Fig.1).

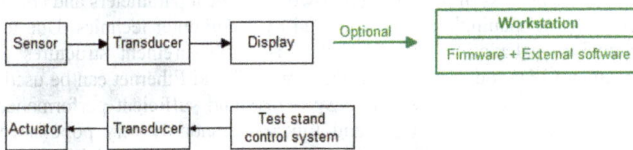

Fig. 1. Traditional measuring device.

The most significant disadvantages of this system are the impossibility to observe non-measurable (indirect) quantities and the need for additional calculations in post-processing; the user has no influence on the structure of the measurement system. By default, the control system of the test stand is independent of the measurement system.

Hence, the main idea for "virtual instruments" is to create a so-called open user interface for software performing data acquisition, processing them according to required algorithms and presenting them in a readable form (e.g., graphs).

Experimental Mechanics   Materials Research Forum LLC
Materials Research Proceedings 30 (2023) 31-38   https://doi.org/10.21741/9781644902578-5

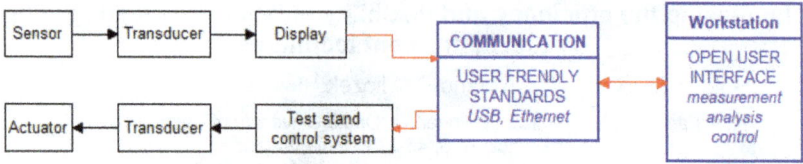

Fig. 2. The idea of virtual instrument.

In this case, actual existing sensors and transducers are used, it is necessary to provide data transfer communication (e.g., USB, Ethernet) and provide a programming environment to create custom programs for data acquisition (DAQ) and analysis. As shown in Fig. 2, it is also possible to integrate measurement and control devices into a single virtual instrument. The flexibility universality of this method favours applications in various fields: mechanical engineering [1, 2], electrical engineering [3] as well as medical science [4, 5], for example. Widespread measurement data transfer (communication) standards make measurement (control) system components easily accessible, where engineers/researchers who are not specialists in the development and programming of measurement systems can carry out integration and IT handling of such systems.

Despite this, most often these DAQ systems are used for basic tasks like monitoring and collecting data from measurement systems. Meanwhile, these environments provide a wide of possibilities for data analysis, from determining parameters by indirect methods and their monitoring during the test, statistical analysis or determination of systems characteristics and also the design of control systems for test equipment [6-10]. For higher expectations in data acquisition and processing, complex so-called real-time techniques are offered. The CompactRIO controllers and field-programmable gate array (FPGA) systems makes it easier for engineers to develop embedded applications designed to control and monitor industrial systems [11, 12]. As it is a hardware component, the system works extremely fast.

The purpose of this overview paper is to discuss the application of the idea of virtual instruments in the development of custom test stands using examples of laboratory tests, where measurements and analysis are performed simultaneously and the data is used in the control process.

Data Acquisition (DAQ) systems

Data acquisition is the process of measuring real-word physical parameters and convert them into digital form that can be manipulated and analysed by a computer technics. Due to the need to program and analyse calculations for data, PC-based measurement structures are found in laboratories. Standard I/O modules like PCI, PCIe or USB and Ethernet can be used. They differ mainly in the speed of data transmission. Easy of use and sufficient performance for typical measurement tasks have made the USB and Ethernet standards very popular recently. One advantage is the mobility of workstations based on laptops, for example. For the applications described in this article, external USB I/O hardware and National Instruments transducers used as:

- NI9205 Module with counter input channels to control the turbine flowmeter,
- NI9237 Strain/Bridge Input Module for handling strain gauge systems and load measurement,
- NI6001 Multifunctional I/O Module, where digital I/O have been used to turn drives on and off and control limit switches, and analog outputs controlled the inverter,
- NI9217 Module for temperature measurement,
- NI9203 Module to serve the absolute pressure transducer, differential pressure transducer, barometer,
- NI9265 Module to control the electric motor rotation,

- NI9403 Module with digital input/output channels to control the stepper motor and limit sensors.

Some of them functioned grouped in a dedicated CompactDAQ cDAQ9172 chassis. The cDAQ system is an extension and facilitation of communication in case of using more measuring DAQ modules.

Standard USB 2.0 Hi Speed is applied, which for small power consumable systems can supply up to 5.25V and 5mA maximum. In this case, it is possible to collect analogue input/output signals with sampling rate 3.2MS/s and update rate 1.6MS/s respectively, digital signal with frequency up to 10MHz.

Virtual instrument programming - graphical programming environment

As mentioned earlier, an essential element of the so-called virtual instruments is an open user interface. Solving the software problem of data acquisition, processing and presentation (e.g., tables, graphs) is the user's task. Custom software allows you to apply your own algorithms for conducting research, processing data, including consideration of physical quantities measured by indirect methods. The difficulty here is knowledge of the programming language. In the case of text-based, highly sophisticated languages like C and its variants, knowledge of syntax, formulating text commands can be very time-consuming and the procedures for communicating with measurement equipment require a lot of experience on the part of the programmer. Supporting engineers and scientists with little programming experience are graphical programming environments. Supporting engineers and scientists with little programming experience are graphical programming environments (e.g. DASYLab, nCode, Matlab SIMULINK or NI LabVIEW). These languages characterized by the fact that the program is created as *a data processing scheme* and built in this form in the workbench (Fig. 3).

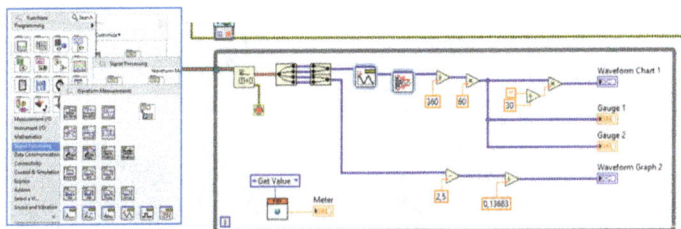

Fig. 3. Functions library (palette) and graphical code of the data flow scheme (NI LabVIEW).

Graphic icons on Figure 3 represent functions and data. They have so-called terminals, to which lines (wires) are connected. The lines represent the flow of data and the direction of flow is strictly defined. A major simplification is defining functions using configuration windows. The user can create his own functions, so-called subroutines. The LabVIEW environment appears to be uniquely advanced in this area: it addresses a variety of engineering issues in measurement programming, data analysis, computer simulation or even machine control systems.

In addition to the graphic code, the user can develop a screen on which the user will communicate with the software (Fig. 4). Again, the user has a library of elements ready to use.

Experimental Mechanics

Materials Research Forum LLC

Materials Research Proceedings 30 (2023) 31-38

https://doi.org/10.21741/9781644902578-5

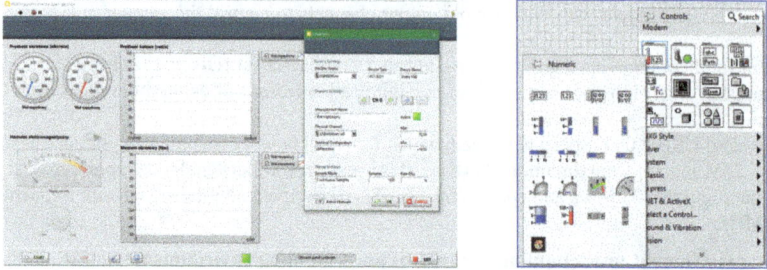

Fig. 4. Front Panel – user interface and screen component library (NI LabVIEW).

Air flow laboratory – case study

The airflow laboratory located at the Opole University of Technology conducts research in the area of phenomena accompanying the flow of air in pipes (Fig. 5a). The size of the gas stream, along with temperature and pressure, are the basic parameters of the operation of such systems. There is a need to study new designs of measuring elements for these phenomena: the presented stand serves, among other things, for this purpose [1, 8, 10]. According to the scope of the study, it is necessary to control the current parameters of the airflow in the pipe. For this purpose, the test stand is equipped with suitable sensors such as a temperature sensor (Fig. 5b), a flow meter (Fig. 5c) or a pressure sensor (Fig. 5d) used for indirect measurement of the jet velocity. The tests have been carry out for different levels of airflow speed. This can be achieved by controlling the speed of the blower drive with an electric motor controlled by an inverter.

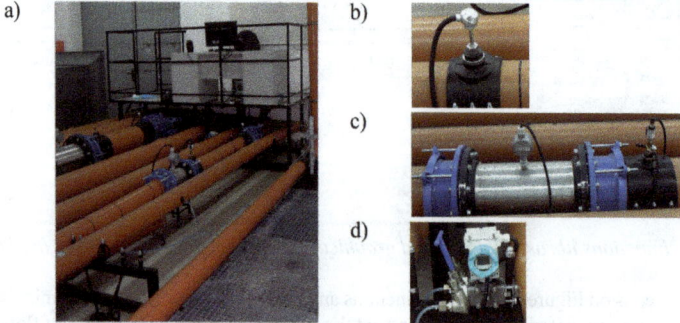

Fig. 5. General view of the stand and example sensors.

In addition, in order for the determination of the air velocity profile in the pipe cross-section to take place automatically the pressure sensor (Prandtl tube) was moved and positioned in the pipe using a linear module with a stepper motor (Fig. 6).

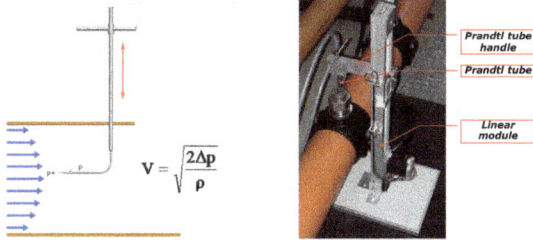

Fig. 6. Measurement of the profile of the air stream velocity.

Both measurements of airflow parameters such as temperature and pressure, positioning of the Prandtl sensor to study the velocity profile, and controlling the blower rotational speed (RPM) have been integrating in a dedicated virtual instrument (Fig. 7).

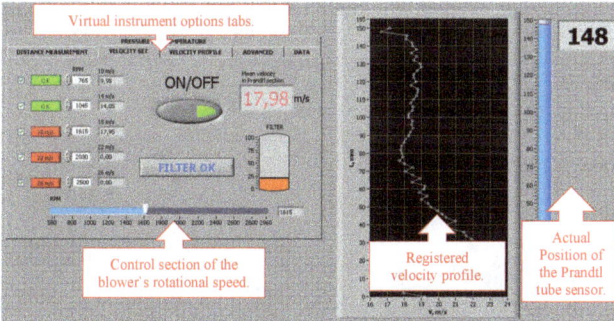

Fig. 7. Virtual instrument for airflow laboratory stand.

As a result, an integrated, completely automatic test and measurement stand has been obtained. The modifications made have resulted in the following functional advantages:
– automatic measurement of the velocity profile of the air stream for different velocities of the jet in the pipe,
– fully monitored and controlled parameters of the air stream in the pipe,
– reduction in the duration of measurements,
– increasing the repeatability of the measurement parameters with controlling and sustaining pre-set values during the test,
– increased resolution of velocity profile measurements.

MZGS fatigue test stand – case study
The MZGS100 fatigue test stand [13] (Fig. 8) was developed for fatigue tests of specimens made from constructional materials subjected to combined bending and torsional loading [14].

The specimen is loaded with moment caused by force applied to the load lever. A rotating disk on which unbalanced weights are placed generates a cyclically time-varying force. As the disk rotates, a centrifugal force B is generated, which, with the help of a link, is transmitted to the beam. An AC motor drives the disk where for rotational speed management the controller with LG iC5

Experimental Mechanics
Materials Research Proceedings 30 (2023) 31-38

Materials Research Forum LLC
https://doi.org/10.21741/9781644902578-5

inverter is applied. An additional spring actuator allows a mean load to be achieved. From a mechanical point of view, the machine is a second-order oscillating object.

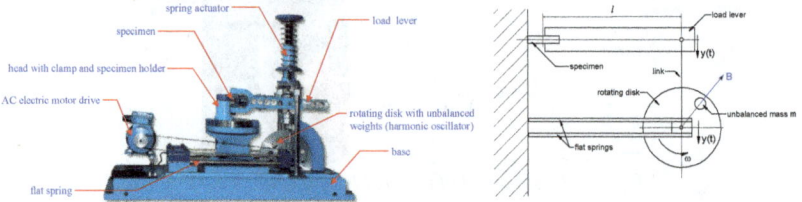

Fig. 8. Overall view of the MZGS100 stand and its scheme [13].

The original design was primarily used to determine the fatigue limit and fatigue properties in the so-called high-cycle fatigue range. Only elastic deformation of the material occurs here, and for proper operation of the stand, it was enough to properly select the weight (base on determined characteristic) and control the speed of the electric motor. The functionality of the machine is strongly limited, and any change in load requires stopping the test and manually changing the operating parameters. Expanding the scope of testing for the area of low-cycle fatigue testing requires controlling test parameters. It is necessary to measure specimen load and displacement. The possibility of changing the load amplitude during operation requires changing the value of the centrifugal force - the speed of the electric motor must be controlled (Fig. 9).

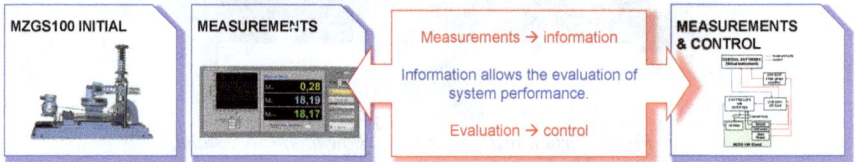

Fig. 9. System development.

Strain gauge sensors act as a load (resultant moment) sensor, elastic deformation of the springs is scaled to measure the displacement of the beam and an encoder integrated with the disk is used to measure the speed of the drive (Fig. 10).
The 4-channel USB 9237 strain gauge amplifier has been use to acquire the signal from the strain gauges. Measurement of the encoder signal was implemented using the counter input of the USB 6001 module. At the same time, the digital outputs of this module controlled the start/stop AC motor connectors of the LG iC5 inverter and the analog output of USB 6001 managed its speed control connectors. Figure 11 shows an example of the screen of the measurement section of a dedicated measuring virtual instrument and an excerpt from the graphical code (schematic) of data processing. During the measurement, it is possible to monitor the accuracy of the drive operation and the current state of the test parameters, observe the time waveforms of load and deformation, the graph of the so-called hysteresis loop is presented as an indirect diagram from the measurement results. Information about test parameters, fragments of data history are saved in the local disk automatically.

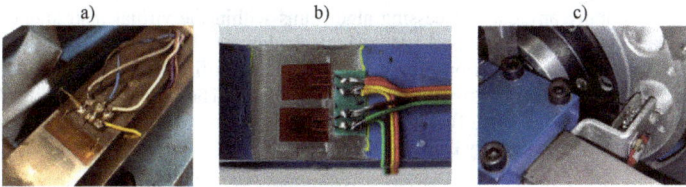

Fig. 10. Sensors: Strain gauges a) on the spring, b) on the beam, c) rotary encoder.

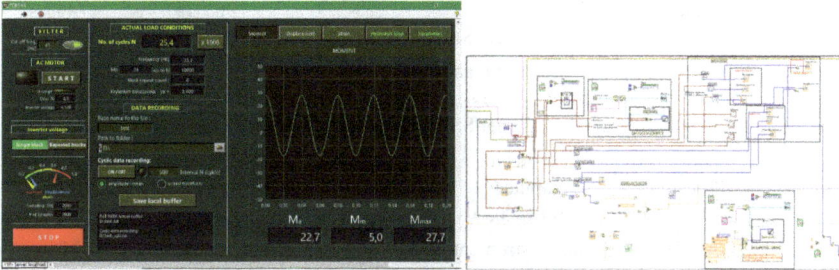

Fig. 11. Virtual instrument for MZGS100 control: measurement module and example of graphical code of the program.

The modifications made have resulted in the following functional advantages:

- configuration of sensors and measurement modules at the control program level,
- full management and control of the parameters of the fatigue test and their sustainment over time,
- performing fatigue tests for both load amplitude and displacement amplitude control,
- automatic load change during tests according to a pre-set sequence of moment amplitude (block load),
- saving data according to user settings,
- control of limit states and automatic shutdown when critical values of operating parameters are exceeded.

Summary

The idea of virtual instrumentation makes available solutions that are effective alternatives to traditional measurement systems. It can be used as a complement to the measurement systems of commercial measurement systems and is an excellent tool for building control and measurement systems for custom laboratory stands. The main advantages of virtual instruments can be drawn as:

- the hardware configuration (e.g. DAQ I/O cards) can support different laboratory stands and tasks as long as the sensor signals meet the data acquisition system's requirements,
- the open user interface allows you to create dedicated programs that process data according to custom (researcher`s) algorithms, especially for physical quantities that are not measurable (indirect quantities),
- both measurements of quantities from the experiment and control of test bench operation can be integrated into the virtual instrument,
- the user can effectively introduce into the existing systems both new measurement equipment and new program elements measurement and control programs,

| Experimental Mechanics | Materials Research Forum LLC |
| Materials Research Proceedings 30 (2023) 31-38 | https://doi.org/10.21741/9781644902578-5 |

- the ability to apply any data processing algorithms within the virtual instrument allows to analyse data on the fly without post-processing calculations.

The presented examples of applications of virtual instruments confirm the possibility of their wider use than just for data acquisition from measuring instruments. The virtual instrumentation based laboratory equipment present cost-effective, compact and user-friendly human-machine interfaces for the measurement and laboratory equipment control.

References

[1] M. Kabaciński, R. Pawliczek, Mechatronic concept for airflow test laboratory equipment, Solid State Phenomena, 220-221 (2015) 445-450. https://doi.org/10.4028/www.scientific.net/SSP.220-221.445

[2] R. Pawliczek, P. Soppa, Measurement and control system for analysis of the operation of the stepper motor, Solid State Phenomena 260 (2017) 113-126. https://doi.org/10.4028/www.scientific.net/SSP.260.113

[3] X. Pei, Virtual instrument based on electronic and electrician's experiment teaching laboratory design, Procedia Computer Science 183 (2021) 120-125. https://doi.org/10.1016/j.procs.2021.02.039

[4] I. Beriliu, H. Falota, Virtual Instrumentation Based Equipment for Bio-medical Measurements, Journal of Electrical and Electronics Engineering, 2 (2009) 155-158.

[5] K. Swain, M. Cherukuri, S.K. Mishra, B. Appasani, S. Patnaik, N. Bizon, LI-Care: A LabVIEW and IoT Based eHealth Monitoring System. Electronics 10 (2021) 3137 https://doi.org/10.3390/electronics10243137

[6] F.J. Jimenez, A.M. Gonzalez, L. Pardo, M. Vazquez-Rodriguez, P. Ochoa, B. Gonzalez, A Virtual Instrument for Measuring the Piezoelectric Coefficients of a Thin Disc in Radial Resonant Mode, Sensors, 21 (2021) 4107. https://doi.org/10.3390/s21124107

[7] M. Almaged, J .Hale, Virtual Instruments Based Approach to Vibration Monitoring, Processing and Analysis, Int. Journal of Instrumentation and Measurement, 4 (2019) 9-16.

[8] M. Kabaciński, R. Pawliczek, Fully automated system for air velocity profile measurement, The Archive of Mechanical Engineering, 59(4) (2012) 435-451. https://doi.org/10.2478/v10180-012-0023-0

[9] R. Pawliczek, Modernization of the fatigue test stand control system using the idea of a virtual instrument, 15th Int. Conference Mechatronic Systems and Materials (Bialystok, Poland), IEEE Xplore (2020) 1-6. https://doi.org/10.1109/MSM49833.2020.9201702

[10] M. Kabaciński, R. Pawliczek, Investigation of vibration effect in measurement system for air flow phenomena in large pipelines, Measurement Automation Monitoring, 62(3) (2016) 96-99.

[11] G. Rata and M. Rata, A solution for study of PID controllers using cRIO system, 9th Int. Symposium on Advanced Topics in Electrical Engineering (ATEE), Bucharest, Romania, 2015, pp. 121-124. https://doi.org/10.1109/ATEE.2015.7133685

[12] I. Moreno-Garcia et all, Real-Time Monitoring System for a Utility-Scale Photovoltaic Power Plant, Sensors 16 (2016) 770. https://doi.org/10.3390/s16060770

[13] H. Achtelik et al., Non-standard fatigue stands for material testing under bending and torsion loadings, AIP Conference Proceedings, 2029 (2018) 020001. https://doi.org/10.1063/1.5066463

[14] K. Kluger, R. Pawliczek, Assessment of Validity of Selected Criteria of Fatigue Life Prediction, Materials (MDPI), 12(14) (2019) 2310. https://doi.org/10.3390/ma12142310

Experimental Mechanics
Materials Research Proceedings 30 (2023) 39-46

Materials Research Forum LLC
https://doi.org/10.21741/9781644902578-6

Extension of PIV methods

Jerzy Pisarek[1,a*]

[1]Kiedrzyńska 95 m 20, POLAND

[a]jerzy.pisarek@gmail.com

Keywords: PIV, Image Analysis, Differential Correlation, Strobing, Sense of Velocity Vector, Diffraction Image of the Particle, Multiphase Flow, Visualisation of the Speed Field, PIV Errors

Abstract. The process of the measurement of the fluid speed by use of PIV method consists most often of several stages: the insertion of the fluid markers of movement, the stimulation of these markers to lucency, the registration of the image of markers, the quantitative analysis of registered images for the purpose of finding the parameters of their movement. Every one of these stages can be realized in many ways. This creates the possibility of adjustment of the measuring-process to realized exploratory tasks. The present article shows possibilities existing in this area, significantly transcending the offer of producers of the measuring apparatus dominant on the market. Additionally, sources of measuring errors ignored usually in manuals of commercial equipment sets are evidenced.

Introduction

Acronym PIV means Particle Image Velocimetry. In the classical version of the method small shiny particles, e.g. solid phase particles, are inserted into the fluid. These particles move together with the fluid, constituting the marker of the movement. Two (or more than two) photographs of the particles images registered in the determined time interval (fig. 1a) enable to determine the speed field in considered area.

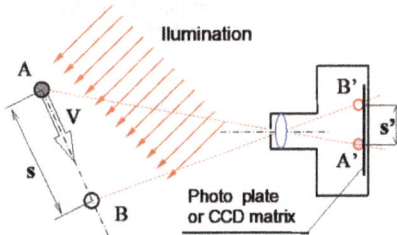

Fig. 1a. Registration of particle image

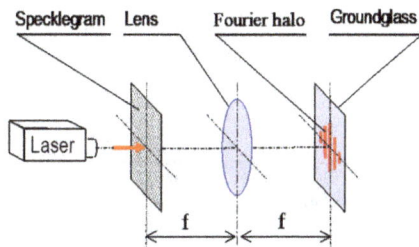

Fig. 1b. Analysis of registered image by use of point by point optical Fourier processor

There are many different kinds of particles, different manners of excitation them to lucency, different techniques of registration of the particle image and different manners of the analysis of image displacement. Classification of certain variants of PIV method elements is presented in table 1.

Experimental Mechanics | Materials Research Forum LLC
Materials Research Proceedings 30 (2023) 39-46 | https://doi.org/10.21741/9781644902578-6

Tab.1. Multiple variants of components of PIV measuring process

	Element of PIV Method	Variants	Marginalia and remarks
1	Marker	Solid body particle: balsa dust, microbalon, flaky aluminum, toner of laser printer Aerosol (in gas media) bubbles of gas in liquid bubbles of gas in gas visible in extended laser beam or in Talbot interferometer	In great deal of commercial offers of PIV arrangement a oil aerosol is very often used as the marker
2	Kind of illumination	Frontal lighting: coherent, noncoherent Lighting from behind: coherent or not coherent Monochromatic optical knife: coherent, noncoherent, Multicolour optical knife Photorescence or luminescence High temperature lighting	Optical knife based on laser light is most often used. Variant "c" in connection with strobing and variant "d" give possible to determine component of speed perpendicular to plane of the light knife.
3	Kind of registration	On photoplate or on CCD matrix Holographic By use of classical single aperture lens By use an photo-objective with stationary or rotating multihole apertures In different interferometer systems	Almost all equipment for PIV used a CCD (charge coup-led device) matrix. The 3D holographic registration was used in TU of Częstochowa by J.Szafrański.
4	Exposure	Registration on following different frames Two short exposures on one frame Multiexposures on one frame Strobing Exposure in with strongly controlled time Time averaging Amplitude or colour modulated impulse	In commercial applications variant "a)" dominates. Time to time variant "b" is used. Variant "g" was developed in TU of Częstochowa by J.Pisarek, A.Wojciechowski, and P.Mirek
5	Whole field image analysis	Many kinds of analog whole field Fourier processors Application of moire effect Digital Fourier processing, binary filtering and interference of images	Whole field analysis is for today used in to advertising and educational purposes
6	Point by point image analysis	Application of point by point optical point Fourier processor Classical digital correlation and autocor-relation Differential correlation Approximation methods, direct determi-nation of components of the relative movements tensor	Techniques based on integral transformations (variant b and c) can be applied only for great quantity of markers in analysed area.

There exist several hundred different configurations of elements constituting the measurement process by use of PIV methods. Unfortunately only several are applied in practice. This results from the strong tendency to the application only of commercial sets of measuring arrangements and proposed by their producers methods of experimental research. The knowledge of persons working in area of the experimental mechanics of fluids is very often restricted to commercial offers of several domineered market companies. It determines the exploratory possibilities in essential way. For purpose of presented article is visible, that the possibility of the significantly better adjustment of the applied methodology of the measurement to the kind of realized tasks

Experimental Mechanics
Materials Research Proceedings 30 (2023) 39-46

Materials Research Forum LLC
https://doi.org/10.21741/9781644902578-6

exists. It is necessary reaching to wider knowledge about the theory of experiment and exit out solutions proposed by producers and sales agents of the measuring equipment.

Historic conditionings

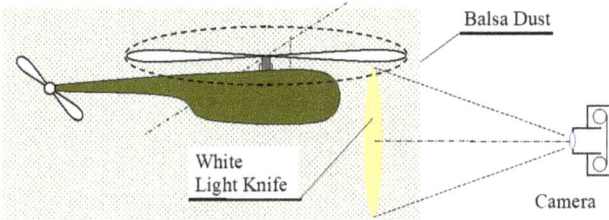

Fig 2. Investigation of the real flow inducted by helicopter rotor, described in NACA reports

The first information of PIV application comes from 50-th years of XX century. As the indicator of the movement of powdered wood of the balsa was applied. Suitably formed geometrically and amplitudely modulated beam from antiaircraft searchlight was a source of the light . Images of dust particles on the classical film plate were registered. In these times presently applied methods of image analysis were still not known. Presently applied phraseology was not used too. Registered film cadres were analysed manually. The army was not interested in a publication of its results in the commercial scientific literature. Therefore the world of the commercial knowledge of new measuring techniques did not notice this new measuring techniques. The dam of the ignorance was broken in the year 1982 when E.Brnaben, J.C.Amare, M.P.Anogo [1] used techniques PIV in laboratory-research and made available results of their works in the English-speaking, high-publishing journal.

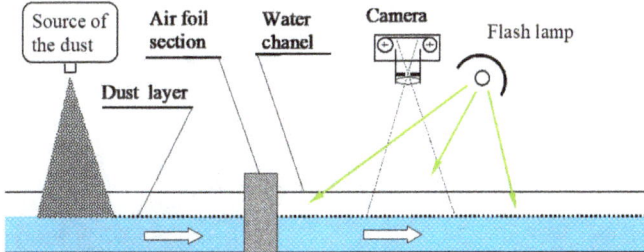

Fig.3. The use of the PIV technique to the investigation of flat flows in the open channel.

As the marker of the movement the aluminum-dust which remained on the water surface as the result of the surface tension was applied. Of course also markers from material with specific gravity smaller than the specific gravity of liquid were possible to use. Molecules of the dust were illuminated from above.

Experimental Mechanics
Materials Research Proceedings 30 (2023) 39-46

Materials Research Forum LLC
https://doi.org/10.21741/9781644902578-6

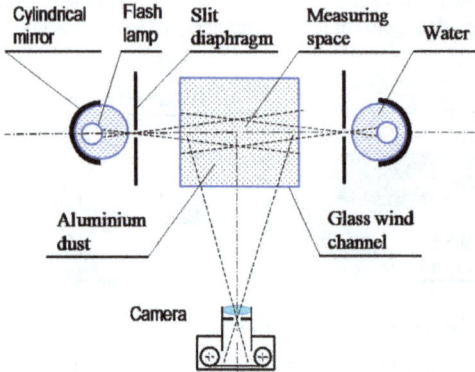

Fig.4. Two white light knife
arrangement for
registration of the air
speed in the wind tunnel

In 1986 J.Pisarek [2] presented the proposal of the measurement of the speed of the fluid in the glass-wind tunnel. The aluminum-pigment was used as the marker. Two light-knives were built on the basis of high voltage bar-flash, applied usually in impulse lasers. In the eighties any descriptions of the equipment PIV the use of laser light knife were presented in many publication 1n year 2001 J.Pisarek and A.Wojciechowski proposed [3] the computerly controlled amplitude modulation of the laser radiation (Fig.5b).

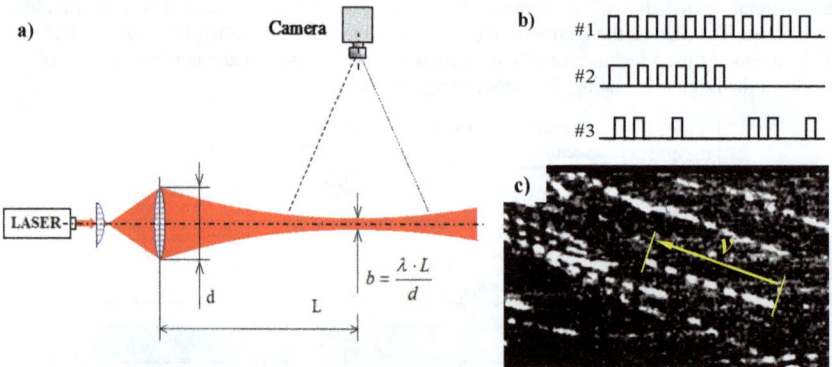

$$b = \frac{\lambda \cdot L}{d}$$

Fig.5. The use of the laser- light-knife with time modulated power

a) the geometry of the knife (where λ is the wavelength of light)
b) examples of graphs of function of light amplitude modulation
c) the image observed at the modulation with the course #2

This gave the possibility of the increase of measuring precision and possibility of designation of sense of a velocity vector on the base of only one film frame. The use of strobing at the large thickness of the light-knife [4,5,6] or at the multicoloured knives gave the possibility of determination of the third constituent of the velocity. The laser beam is a gaussian beam which properties are described in handbooks of the wave-optics [9,10]. The laser optical knife is one of forms of transformation of the gaussian beam. By selection of suitably parameters of the optical system one can, for the determined wavelength, fix the average thickness of the knife and his width and to enumerate changes of these parameters in the area of measuring.

The registration of the picture of particle constituting the marker of the movement surrounding fluid can be made by the camera lens with single-hole or multihole aperture diaphragm.

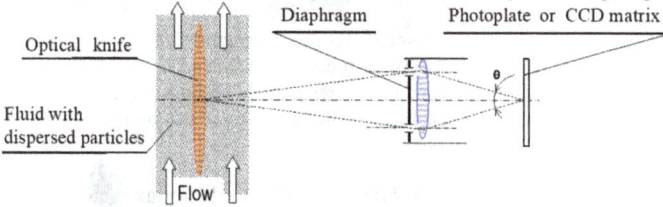

Fig. 6. The registration of particles image by the lens with the two-hole aperture diaphragm

In this second case (fig.6) in the plane of the picture the row of interferential fringes is observed. Fringes are perpendicular to the straight line passing trough holes in the diaphragm and distant each from other is inversely proportional to the length of the distance between these holes. If on material with the strongly non-linear attenuation diagram two different positions of the particle in the small distance will be registered then the interference of images (moire effect) will be observed. The same effect will follow in case of the digital processing of the image registered on linear material transformed in to B&W image. In case of the peck of simultaneously registered particles moving in the same direction with different speeds we will receive the contour-map of the speed component parallel to the line connective diaphragm holes.

Analog Fourier processors
Idea of point by point processor was presented on fig.1b. In practice the optical system can be extended a little more. The theory of optical Fourier transforming is described in the work [10].

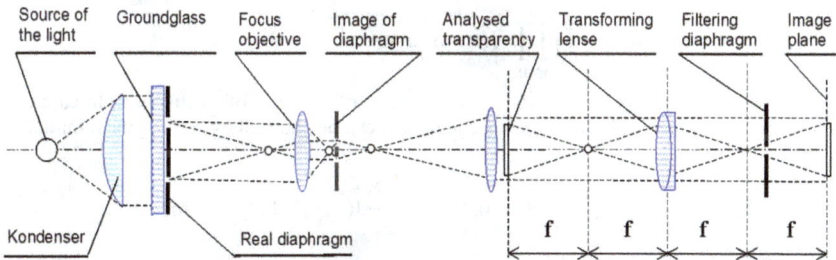

Fig.7. The diagram of optical Fourier processor to determining of the speed module.

Whole field processing of PIV image and the theory of Fourier transformation of multiexposure speckle images, (to which PIV images belong), are described in work [7]. New solution, hitherto not published, is the processor enabling the visualization of the spatial schedule of modules of the speed (fig.7). The light source of containing the diaphragm with at least one annular (ring) hole is a characteristic feature of proposed arrangement. The hole in the diaphragm can be filled by a circular-symmetrical multicoloured filter or to contain a circular-symmetrical grating. The filtration of spatial frequencies of analysed transparency takes place by the small single hole situated centrally in the plane of transforms.

Digital analysis of the image
The algorithm of the quick two-dimensional fast Fourier transformation (FFT) is currently very often used technique of multi-exposure particle image analysis. If in the virtual transform plane the binary filter will be put and the reverse transform of the function modified by this filter will be

made then the contour-map of the determined speed component on whole field measuring region will be possible to obtain. This is the digital implementation of whole field optical Fourier processors.

For point by point analysis of the speed, recorded trough PIV technique the correlation or autocorrelation algorithms can be used. In most cases there are product-algorithms.

Let us assume that two different frames with registered two images are translocated in relation to themselves \by vector $s=[s_x, s_y]$. Function $K_1(u,v)$ described by the equation (1) reach a maximum when $|u|=|s_x|$ and $|v|=|s_y|$.

$$K_1(u,v) = \iint_A J_1(x,y) \cdot J_2(x+u,y+v)\,dxdy = max \qquad (1)$$

when

x,y – Coordinates in plane of analysed image
u,v – Coordinates in virtual plane in space of integral translation
A – Analysed area of the image
J(x,y) – The function of the brightness of points of images registered in the first or second exposure

In case of when both images are registered on the same film frame the functional K2 described by formula 2 achieves the maximum when $|u|=|s_x|$ and $|v|=|s_y|$

$$K_2(u,v) = \iint_A J(x,y) \cdot J(x+u,y+v)\,dxdy = max \qquad (2)$$

In case of registration of N-exposures the absolute value of the shift of analyzed image region one can obtain from the condition of the maximization of the functional value:

$$K_3(u,v) = \iint_A \prod_{n=0}^{N} J(x+n\cdot u, y+n\cdot v)\,dxdy = max \qquad (3)$$

Differential algorithms give a radical improvement of accuracy of calculations. In case of N exposures sought module of movements component can be obtained by seeking the minimum of functional:

$$K_4(u,v) = \sum_{n=1}^{N} \iint_A \left(J(x+n\cdot u, y+n\cdot v) - J(x,y)\right)^2 dxdy = min \qquad (4)$$

Given algorithms permit the designation from one film frame of only absolute values of movement components. However the possibility of the designation of the speed sense exists if we apply the suitable modulation of laser pulse. For example for modulation #3 from fig.3b the sense of the velocity vector one can obtain designating the maximum of the function:

$$K_5(u,v) = \iint_A \left(\left(J(x,y)\right) \cdot \left(J(x+u,y+v)\right) \cdot \left(J(x+3u,y+3v)\right)\right)\,dxdy = max \qquad (5)$$

Described higher algorithms are intended to the analysis of the speed of the fluid in which the large volumetric concentration of markers is observed. In case of small concentrations of particles and in case of some kinds of multi-phase flows one ought to apply algorithms basing an analysis of trajectory given out particle. The use of the theory of cliques gives here good results.

Experimental Mechanics
Materials Research Proceedings 30 (2023) 39-46

Materials Research Forum LLC
https://doi.org/10.21741/9781644902578-6

Atypical uses

A spectacular example of the use of the technique PIV is technique of analysis of traffic of solid body particles in thick fluidal layer, elaborated in TU Częstochowa by K.Sikora and J.Pisarek. The modelling of the fluidal-layer by the set of identical particles with a very dark or

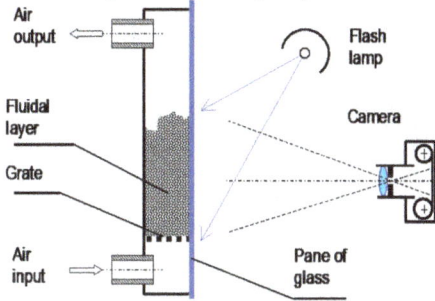

Fig.8a. Configuration of statement for measuring of *processes in high dense fluidal layer*

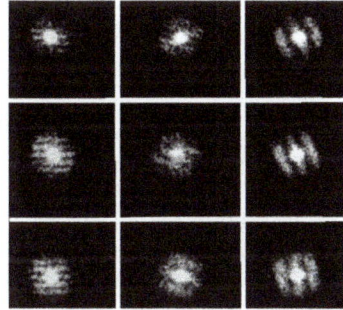

Fig.8b. Fourier halos obtained in optical point by point processor from transparency recorded in arrangement shown on fig. 8a

very bright colour is a general idea of this technique. Number of bright particles should be at most 5-times smaller than number of dark particles. The diagram of the measuring-arrangement one showed on fig.8a. The example of result of the analysis in point by point Fourier processor is showed on fig.8.b.

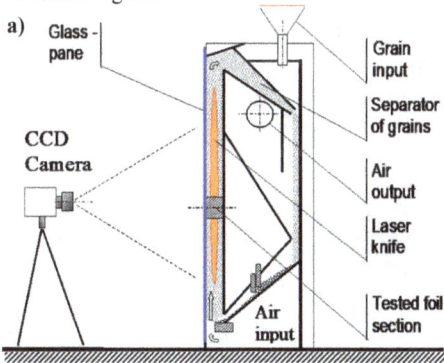

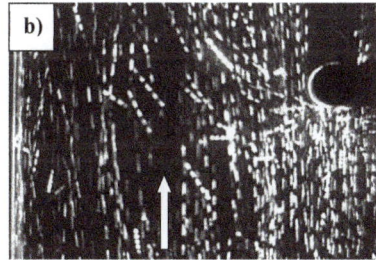

Fig. 9.
a) *Desk table wind table tunnel*
b) *The fragment of typical image registered from this tunnel*

The PIV technique can be applied also to model investigation of circulation fluidal layers and to the illustration of some phenomena of the mechanics of fluids. The transportable arrangement is made from glass and has dimensions 1000x400x200 mm. The schema of tunnel is shown on fig.9a. The air flow was forced by usual vacuum cleaner. As particles of the solid phase alternatively: the table salt or sand (for two phase flow) , the semolina (for gas flow visualization) were applied.

Disadvantages and errors

The basic disadvantages of PIV methods is the lack of the possibility of making a measurement in real-time. The inertial reaction and the gravitation can influence on the particle equally strongly as surrounding her gas. Additionally electrostatic forces can be significantly sources of measuring errors. The movement of speckle pattern generated in laser light by the gyral solid body particle is also often observed source of error.

References

[1] E.Brnaben, J.C.Amare, M.P.Anogo – „White light speckle method for measurement of flow velocity fields" – Appl.Opt., Vol.21, № 19, 1982 , p. 3520-3527. https://doi.org/10.1364/AO.21.002583

[2] J.Pisarek – „Doświadczalna analiza odkształceń i przemieszczeń metodami fotografii plamkowej w świetle białym" – rozprawa doktorska, Politechnika Częstochowska luty 1986

[3] J.Pisarek, A.Wojciechowski – „Pomiar prędkości płynu i cząstek stałych zmodyfikowaną metodą PIT " – materiały sympozjum Optoelektronika 2001, Warszawa 2001, str.173

[4] W.Nowak, J.Pisarek, P.Mirek – „Application of Laser Sheet Technique for Analysis of 3D Particle Velocity Fields in Fluidized Beds. Fluidisation (I)" – Vol. 14, 2000 No 75 p. 355-363

[5] P. Mirek –„Application of Laser Sheet Technique for Analysis of 3D Particle Velocity Fields" – Journal of Theoretical and Applied Mechanics, 1, 39, 2001, p. 33-49

[6] P.Mirek – „Laserometryczne metody pomiaru przepływów dwufazowych ziarna-gaz" – Dissertation , Częstochowa 2002

[7] J.Pisarek – „Оптико-цифрові методі і системі аналізу спеклограм для визначення полів переміщень і деформацій" – hab. Lwów 1996

[8] R.J.Adrian, J.Westerweel – „Partickle Image Velocimetry" Cambridge 2011

[9] R.Jóźwicki – „Optyka laserów" – WNT 1981

[10] W.T.Cathey – „Optyczne przetwarzanie informacji i holografia" – PWN 1978

Experimental Mechanics
Materials Research Proceedings 30 (2023) 47-54

Materials Research Forum LLC
https://doi.org/10.21741/9781644902578-7

Analytical, numerical and bench tests of axles in rail vehicles

Patrycja Lau[1,2,a*], Piotr Paczos[2,b]

[1] Lukasiewicz Research Network - Poznań Institute of Technology, Rail Vehicles Center, 181 Warszawska Street, 61-055 Poznań, Poland

[2] Poznan University of Technology, Institute of Applied Mechanics, Plac Marii Skłodowskiej-Curie 5, 60-965 Poznań, Poland

[a] patrycja.lau@pit.lukasiewicz.gov.pl, [b] piotr.paczos@put.poznan.pl

Keywords: Wheelset Axles, Fatigue Strength, Experimental Tests, Alternative Method, Rail Vehicles, FEM Calculations, Analytical Research

Abstract. On the basis of experimental, analytical and numerical tests, a strength analysis of a rail vehicle axle was presented, as well as an alternative approach to this type of issue. The axle of the wheelset was tested on the experimental stand. Then, analytical calculations of the tested axis were performed in accordance with the EN 13 103-1 standard in the sections where strain gauges were located during the stand test. A numerical model was also created in a program based on the Finite Element Method. The obtained results were compared and summarized. It turned out that the results from all studies coincided, which suggests that each of the methods used is correct. None of the obtained values exceeded the permissible fatigue stresses.

Introduction

Wheelsets (axle + wheels) are the basic construction unit of a rail vehicle. Among many assemblies, they are the most exposed to fatigue wear, which was also strongly emphasized in articles [1, 2]. Sobaś [3] presented technological measures increasing the service life of wheelset axles on the basis of applicable standards. Michnej and Krwala [4] characterized the surface-reinforcement treatments that increase the durability of the axles of railway wheelsets. Whereas, Antolik [5] described the sources of fatigue cracks in railway axles. Many studies on the durability of railway axles and attempts to strengthen the axles show the essence of the problem. This is particularly important in the currently designed vehicles, where the main assumptions are to design a vehicle with the lowest possible weight, moving at higher and higher cruising speeds.

The current applicable European standard EN 13 103-1 [6] containing the necessary rules for the construction and testing of wheelset axles clearly suggests that axle strength calculations should be performed using the analytical method. Thanks to the rules contained there, it is possible to correctly perform a mathematical model of the tested object and correctly determine the forces and boundary conditions for analytical calculations. Nevertheless, in this work, in order to compare the analytical method and experimental tests with FEM simulation, it is necessary to thoroughly understand the operation of the machine for testing the fatigue strength of axles of wheelsets, which was presented in detail in Stasiak's academic textbook [7].

Nowadays, analytical methods are often superseded or only supplement the finite element method, which is more accurate and has a wider range of applications. Similarly, in the case of axis calculations, it seems necessary to adapt the current approach to modern simulation tools. An alternative axis calculation method would be a bridge between analytical calculations resulting from the standard and simulation methods, which would significantly shorten the calculation time and enable multi-directional analyses.

Experimental Mechanics
Materials Research Proceedings 30 (2023) 47-54

Materials Research Forum LLC
https://doi.org/10.21741/9781644902578-7

Tested axle

The object of research of this work is the non-powered axle of the type A wheelset (PN-92/K-91048) [8]. The axle is made of EA1N steel.

The values of permissible stresses are presented in table 1. The values have been selected according to the European standard EN 13103-1 [6] and they result from the fatigue limit at rotational bending for the axis. It takes into account the safety factor S = 1.2 and the fact that the places where we will make the measurement are outside the embedment areas.

Table 1 Maximum permissible stresses for solid axles [6]

Steel	σ_{dop}	σ_{dop} S=1.2
	[MPa]	[MPa]
EA1N	200	166

Bench tests

Bench tests were carried out at the Łukasiewicz Research Network - Poznań Institute of Technology in the Rail Vehicles Testing Laboratory at the 18SB test stand intended for testing the fatigue strength of wheelsets axles. For this purpose, a sample was delivered to the plant in the form of a half-set, which consisted of an axle and one pressed-in wheel. The wheel was pressed onto the axle with the applicable dimensional tolerances and forces. The tests included the performance of a fatigue test in the range of 10 million cycles and strain gauge measurements to determine the stresses and control of these loads during the test. Strain gauge measurements belong to experimental methods of measuring deformations on the surface of the tested element. The centers of the electro-resistance strain gauges were respectively 300 and 250 [mm] from the center of the wheel (Fig. 1).

Fig. 1 View of the arrangement of resistance strain gauges

Experimental Mechanics

Materials Research Proceedings 30 (2023) 47-54

Materials Research Forum LLC

https://doi.org/10.21741/9781644902578-7

The structural system of the fatigue machine used for testing, shown in figure 2, uses the centrifugal forces of masses rotating with a constant angular velocity. A sample of the axle with actual dimensions was attached vertically to the machine body, and the rolling wheel mounted on the axle was attached with special anchor holders. Forces P_1 and P_2 (values: 37.4 and 81.0 [kN]), which directly loaded the sample, were caused by centrifugal pulsators mounted at its ends. The pulsators that caused the Q_1 and Q_2 forces were used to dynamically balance the forces that acted on the foundation of the machine. Both the upper and the lower pulsator are driven by a separate DC motor, powered by a common drive system from the control panel. In the middle part of the column there were safety sensors, which, in the event of contact with the axle sample, immediately switched off the drive system and stopped the machine. The machine was made at H. Cegielski in Poznań according to the construction documentation of the Research and Development Center of Rail Vehicles (now Łukasiewicz – PIT).

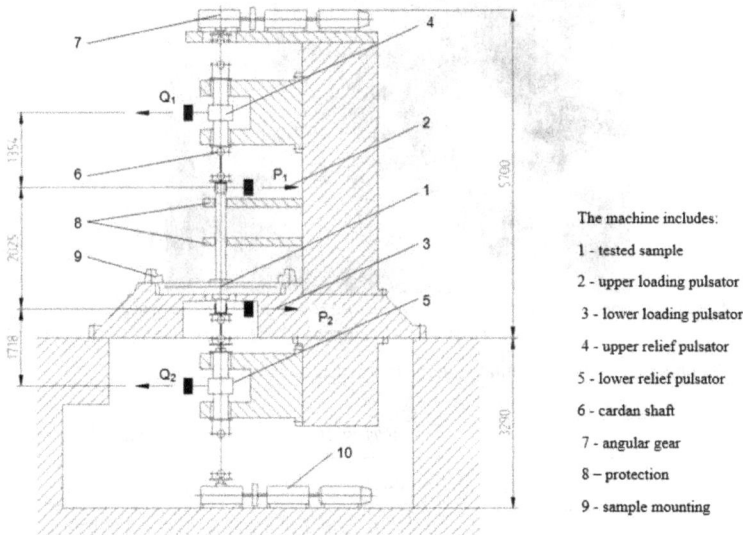

The machine includes:

1 - tested sample

2 - upper loading pulsator

3 - lower loading pulsator

4 - upper relief pulsator

5 - lower relief pulsator

6 - cardan shaft

7 - angular gear

8 – protection

9 - sample mounting

Fig. 2 Scheme of a fatigue machine for testing axles and wheels of wheelsets [7]

The tested set was mounted on the fatigue machine shown in figure 3 and statically loaded in the axis of the upper pulsator with a concentrated force corresponding to the centrifugal force of the pulsator. This procedure is necessary to determine the important parameters for setting the loads required in the fatigue test. During two measurement cycles, the axes were loaded with a force of 40 kN and the deflection arrow was measured in the axis of the upper pulsator and in the axis of the safety sensor, and strain gauges were measured in the control zones. In this way, the rotating masses of the loading pulsators were determined.

Experimental Mechanics

Materials Research Proceedings 30 (2023) 47-54

Materials Research Forum LLC

https://doi.org/10.21741/9781644902578-7

Fig. 3 View of the fatigue machine for testing axles and wheels of wheelsets (Łukasiewicz-PIT archive)

The results obtained in experimental studies are presented in table 2.

Table 2 Deformations and stresses occurring at the locations of strain gauges

Strain gauge no	Distance	Deformation		Average amplitude	
		ε_{max}	ε_{max}	$\varepsilon_{śr}$	σ_a
	[mm]	[μm/m]	[μm/m]	[μm/m]	[MPa]
1 os	300	585.5	596.05	590.775	122
2 os	250	608.37	619.48	613.925	127

Analytical calculations

Analytical tests were carried out in accordance with the European standard EN 13103-1 [6]. As a standard, calculations are made considering two types of forces coming from masses in motion and braking. The concentration of stresses in the cross-sections of the axles most exposed to overloads is checked. Nevertheless, the purposes of comparison with other test methods, two cross-sections in which the strain gauges were placed during the bench test were analyzed analytically.

Experimental Mechanics   Materials Research Forum LLC
Materials Research Proceedings 30 (2023) 47-54   https://doi.org/10.21741/9781644902578-7

The input data needed for the calculations are given in Table 3.

Table 3 Results

Force on the upper pulsator	P_1	37.40	kN
Force on the lower pulsator	P_2	-81.00	kN
Wheel rolling radius	R	0.46	m
Distance of the upper pulsator axis to the center of the wheel	L_g	1.74	m
Distance of the lower pulsator axis to the center of the wheel	L_d	0.26	m
Top moment at the center of the rim width	$M_g = P_1 \cdot L_g$	65.15	kN·m
Bottom moment at the center of the rim width	$M_d = P_2 \cdot L_d$	-20.90	kN·m
Moment applied to the wheel disc M_1-M_2	$M_t = M_g - M_d$	86.05	kN·m
Force at the periphery of the circle	$P_t = \dfrac{M_t}{R}$	187.06	kN
Distance of the wheel from the „2os" strain gauge on the axle	L_{i1}	0.25	m
Distance of the wheel from the „1os" strain gauge on the axle	L_{i2}	0.30	m
Axle dimeter behind the wheel	d	0.160	m
Strength index	$W = \pi \cdot \dfrac{d^3}{32}$	0.000402	m^3
The bending moment of the axis in the place of the strain gauge „2os"	$M_1 = P_1(L_g - L_{i1})$	55.80	kN·m
The bending moment of the axis in the place of the strain gauge „1os"	$M_2 = P_1(L_g - L_{i2})$	53.93	kN·m
Stresses in the axis at the place strain gauge „2os" <166	$S_1 = \dfrac{M_1}{W \cdot 1000}$	139	MPa
Stresses in the axis at the place strain gauge „1os" <166	$S_2 = \dfrac{M_2}{W \cdot 1000}$	134	MPa
The ratio of the moments Mg/Md	t	-3.12	-

MES research

Currently, the Finite Element Method is the most used tool for numerical analysis of structures. It is a discrete method that solves differential equations in an approximate way. The use of FEM allows for significant savings by eliminating many bench tests. Thanks to it, it is also possible to properly prepare the model for experimental research by determining the appropriate measurement parameters or indicating the most favorable location of strain gauges.

The strength analysis was performed with the ABAQUS/Standard program based on the Finite Element Method at the Łukasiewicz Research Network - Poznań Institute of Technology.

The following units of measurement were used in the calculations: [mm], [N], [MPa].

Support conditions and loads P_1 and P_2 (Fig. 4) were introduced into the computational model, which are identical to the restraint of the wheelset during the bench test and the assumptions that

Experimental Mechanics

Materials Research Proceedings 30 (2023) 47-54

Materials Research Forum LLC

https://doi.org/10.21741/9781644902578-7

were included in the mathematical model, so that the obtained results can be compared as precisely as possible from all conducted studies.

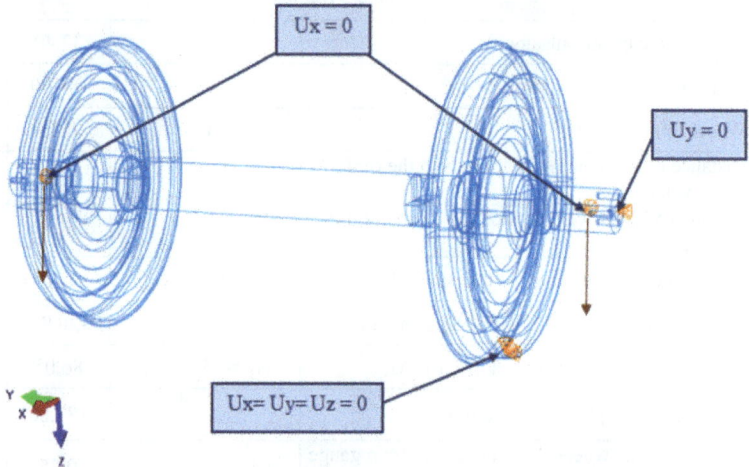

Fig. 4 Scheme of modeling the boundary conditions of the tested object

Before performing the calculations, a study of the size and type of the finite element was carried out. The stress calculation results were examined and compared by changing the types of finite elements on the axis. Various finite elements were checked and the results of the calculations turned out to be were very similar to each other. It was decided to use tetragonal square elements in accordance with the guidelines in [9]. This allows to reproduce all the details of the tested axis with the greatest possible accuracy and to easily determine the points where the strain gauges were placed during stand tests. The results obtained with the FEM method are shown in figure 5.

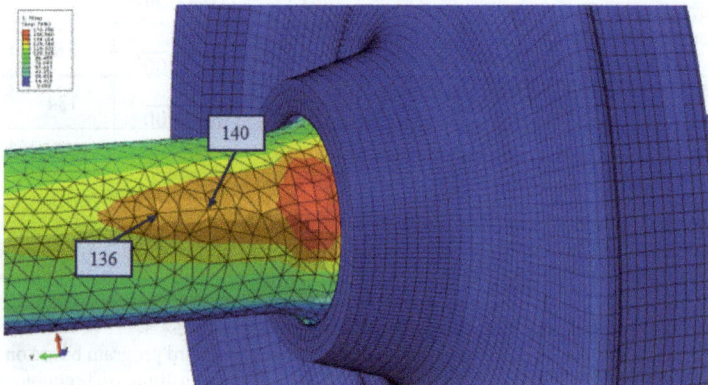

Fig. 5 Huber-Mises stress distribution in the wheelset model in the locations of strain gauges

Experimental Mechanics

Materials Research Proceedings 30 (2023) 47-54

Materials Research Forum LLC

https://doi.org/10.21741/9781644902578-7

Summary

Comparison of the results obtained from analytical calculations, FEM numerical tests and bench tests with limit values is presented in Table 4.

Table 4 Summary of research results

Strain gauge no	Distance	Bench research	Calculations analytical	FEM numerical research	Permissible stresses [6]
		σ_a	σ_a	σ_a	σ_{dop}
	[mm]	[MPa]	[MPa]	[MPa]	[MPa]
1 os	300	124	134	136	166
2 os	250	129	139	140	166

The stress values obtained from analytical calculations of the tested axle of the rail vehicle and during numerical tests based on the Finite Element Method, as well as bench tests, are almost identical, which suggests the correctness of determining the boundary conditions and loads in computer analysis using the Abaqus/Standard program. None of the obtained values exceeded the permissible stress (taking into account the safety factor equal to 1.2), i.e. 166 MPa.

Thanks to the conducted research, it can be concluded that numerical analysis based on the finite element method reflect the real results that were obtained by means of deformations determined on the tested axle sample on the experimental stand and during analytical calculations in accordance with the standard [6].

On the basis of the conducted research, it can be unequivocally stated that simulation studies can be an alternative or a great support for analytical methods. Thanks to the FEM analysis, we can check the distribution of stresses that arises over the entire tested structure, and not only in a few potentially most endangered cross-sections. Using computer simulation, we can additionally introduce various types of structural modifications much faster and check how a given change will affect the strength of the tested object. In many areas, the finite element method has completely supplanted analytical calculations. Also, in the case of axles of wheelsets in rail vehicles, it seems necessary, for example, to extend the existing guidelines for axle fatigue strength analysis with FEM simulations. The discussed extension of the strength analysis will significantly facilitate and accelerate the work on the design of the axles of rail vehicles and will enrich the methodology of approach to such issues.

To enrich the strength analysis of rail vehicle axles carried out so far, will be to carry out bench tests with more strain gauges. It gives us a more accurate picture of the stress distribution in the real axle located on the experimental stand and compare it with the results of analytical calculations and FEM simulations.

Acknowledgments

The research was carried out as part of the Implementation Doctorate Program of the Ministry of Education and Science implemented in the years 2021-2025 (Contract no DWD/5/0378/2021)

References

[1] S. Kowalski, Selected problems in the exploitation of wheelsets in rail vehicles. Journal of Machine Construction and Maintenance 2/2017

[2] C. Song, M.X. Shen, X.F. Lin, D.W. Li, M.H. Zhu, An investigation on rotatory bending fretting fatigue damage of railway axles. Fatigue Fract Engng Mater Struct, 2014. https://doi.org/10.1111/ffe.12085

[3] M. Sobaś, Technological treatments to increase projected axle life of the wheelsets, Rail Vehicles 4/2011 (In Polish)

[4] M. Michnej, K. Krwala, Technological methods of increasing durability of the axles of rail vehicle wheelsets. Logistyka 3/2015 (In Polish)

[5] Ł. Antolik, Methodology of detecting fatigue cracks in railway axles and the requirements of European standards, Problemy Kolejnictwa, 165, December 2014 (In Polish)

[6] EN 13103-1 Railway applications - Wheelsets and bogies - Part 1: Design method for axles with external journals

[7] L. Stasiak, Experimental determination of the fatigue strength characteristics of the wheelset axles of rail vehicles, Paper Nr 173, Poznań University of Technology, Poznań 1986 (In Polish)

[8] Polish standard PN-92/K-91048

[9] E. Wang, N. Thomas, R. Rainer, Back to elements - Tetrahedra vs. Hexahedra. CAD-FEM GmbH, Germany 2004

Experimental Mechanics
Materials Research Proceedings 30 (2023) 55-60

Materials Research Forum LLC
https://doi.org/10.21741/9781644902578-8

Investigation of electromechanical coupling characteristics of a double magnet system

Andrzej Mitura[1,a *] and Krzysztof Kecik[1,b]

[1]Lublin University of Technology, Faculty of Mechanical Engineering, Department of Applied Mechanics, Nadbystrzycka 36, Lublin, Poland

[a]a.mitura@pollub.pl, [b]k.kecik@pollub.pl

Keywords: Electromechanical Coupling, Energy Harvesting, Superposition

Abstract. In this work, the experimental results of the electromechanical coupling coefficient identification are presented. The research covers two cases: test with a single magnet (I) and test with a double magnet, where two repelling magnets in one structure are connected (II). In case (I), the analytical description of the electromechanical coupling coefficient was determined. Whereas the analysis of case (II) confirms the superposition principle for interactions between both magnets and a single inductive coil. Finally, the presented analysis proposes some premises which will be used in the future to develop the model with two levitating magnets.

Introduction

Energy harvesting from a system with one levitating magnet is well-known and has been described in many papers [1, 2, 3]. This research presents many aspects: the modelling of electromechanical coupling [1], the existence of electrical damping [2], optimal resistance load [3], etc. However, one of the problems of this solution is its low recovery effectiveness. For example, a small electromagnetic harvester dedicated to energy recovery from human body motion is presented in [4]. Obtained results showed that the maximum harvested electrical power was approx. 0.7mW. In order to increase efficiency, modifications to the electromagnetic harvester structure are introduced. One of the suggestions is to use two independent levitating magnets. This concept is presented by Mitura and Kecik in [5]. This research has several limitations. First, the electromechanical coupling coefficient has a constant value, because small vibrations from magnets are only considered. Secondly, each magnet has its inductive coil and the determination of the total electrical power is ambiguous. However, the results obtained are promising. The maximum recovered electrical power was about 0.45W. In the next research, it is planned to use only one coil to induce an electromotive force from the larger vibrations of both movable magnets. Developing a new model requires describing the strongly nonlinear curve of the electromechanical coupling coefficient. Moreover, the mechanism of the electromotive force generation should be checked, i.e. is it a superposition of the interactions between individual magnets and a single coil.

The research shown in this paper will be used in the future to develop a model with two levitating magnets and a single inductive coil. The presented analysis is limited to two selected issues from the initial modelling stage (sections 2 and 3). In the section "Experimental research with a single magnet" the experimental measurement of the electromechanical coupling coefficient for the motion of a single magnet relative to the coil is presented. The obtained nonlinear relationship is approximated by analytic functions. This description can be used to model the electromechanical coupling coefficient of interaction between one magnet and a coil. The next section "Experimental research with double magnet" presents experimental research, where two moving magnets were used. The purpose of this study was to show that descriptions of the interaction between each magnet and the coil can be considered separately. Tests for two independent magnets would be inconclusive. Therefore, the special case was analyzed. In this case, both magnets were connected by a connector. Finally, the distance between both magnets is

Experimental Mechanics   Materials Research Forum LLC
Materials Research Proceedings 30 (2023) 55-60   https://doi.org/10.21741/9781644902578-8

constant. The created structure (magnet-connector-magnet) called a double magnet is easier to analyze, and the superposition principle could be confirmed. The superposition principle means that the coil voltage obtained from the double magnet motion results from the sum of the voltages generated by two single magnets considered separately. In the last section, the conclusions are presented.

Experimental research with a single magnet

All experimental research was performed on a small strength machine, Shimadzu. An epoxy tube was clamped in the bottom fixed machine handle. Next, an inductive coil was installed on this tube. The basic parameters of the applied ring-shaped coil were: total length - 50 mm, inner diameter - 28 mm, outer diameter - 42 mm, resistance - 1.15 kΩ and inductance - 1.46 H. The tube also provided proper guidance of the neodymium magnet relative to the coil. The cylindrical magnet used was 20 mm in diameter and 20 mm in height and was connected to the machine traverse. During tests, this traverse moved up and down at a constant speed (1 meter per 1 minute) and generates the motion of the magnet relative to the coil. Based on the experiment, the electromotive force U (coil voltage in volts) and magnet position relative to the coil centre x (distance in millimetres) were measured. Obtained nonlinear curve $U=f(x)$ is presented in Fig.1a. These data were used for the determination of electromechanical coupling coefficient α. Simple calculations can be made from the following equation:

$$\alpha=U/v, \tag{1}$$

where v is the traverse speed (0.0167 m/s). In Fig. 1b the nonlinear characteristics $\alpha=f(x)$ can be seen.

a)   b)

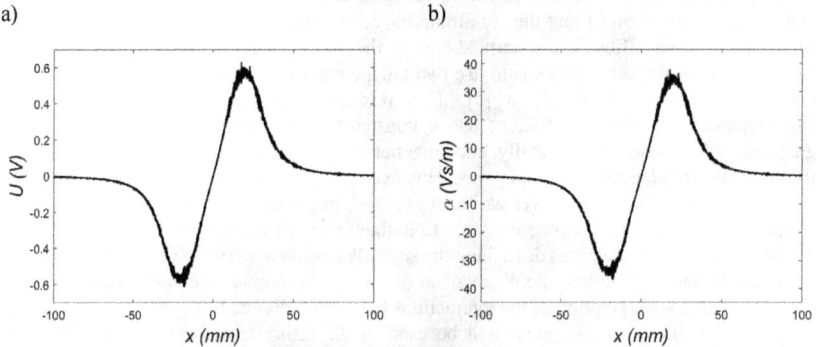

Fig.1. The measured curve $U=f(x)$ (a) and approximation $\alpha=f(x)$ (b) based on it.

The trend $\alpha=f(x)$ can be described by a polynomial [6]:

$$\alpha(x)=\alpha_0+\alpha_1 x+\alpha_2 x^2+\alpha_3 x^3+...+\alpha_n x^n. \tag{2}$$

Polynomial coefficients were found using the polyfit MATLAB function [7]. Quality analysis of the experimental data fitting by n^{th} order polynomial was also performed. The fit quality was assessed using the mean absolute error MAE [8]:

$$\Delta = \frac{1}{m} \sum_{i=1}^{m} \left| \alpha_{i,experiment} - \alpha_{i,polynomial} \right|, \tag{3}$$

where: m is points number, $\alpha_{i,experiment}$ is i^{th} value of electromechanical coupling coefficient calculated from the experiment (1) and $\alpha_{i,polynomial}$ is i^{th} value of α taken from polyfit approximation (2). The influence of order number - n on the fit error is illustrated in Fig.2a. When $n>24$ it can be assumed that the error Δ does not change and it is minimal, about 0.32 Vs/m.

a)   b)

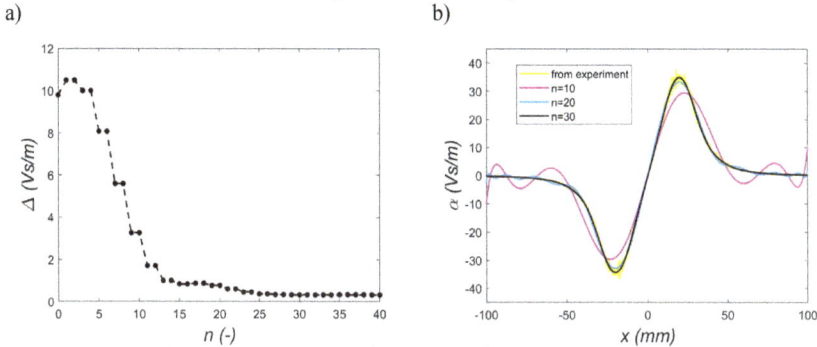

Fig.2. The fit error $\Delta=f(n)$ (a) and selected polynomial approximation $\alpha=f(x)$ (b).

After analyzing the curves in Fig.2b, it can be concluded that a good representation of the experimental data requires the use of a high-order polynomial.

Experimental research with double magnet
In the previous section, the electromechanical coupling coefficient model (2) for interaction between a single magnet and inductive coil is given. If this relationship will be applied to a system with two levitating magnets, then the relationship independence between each magnet and the coil should be checked. In this section, this interactions independence is confirmed. To research, two identical magnets were used. The applied orientation of their poles would repeal magnets (see Fig. 3). So, both magnets were connected with a connector. In this situation, the distance between both magnets is constant. This is also important from an analysis point of view. Now, it is much simpler and unambiguous. The structure of the so-called double magnet (magnet - connector - magnet) is shown in Fig.3. This scheme presents and defines the basic elements (1, 2, 3, 4, 5), centers of connector and coil (points O and C, respectively), lengths of magnet and connector (l - magnet, L-connector), and coordinate x. The experimental tests were repeated for different lengths of the connector L: 0 mm, 15mm, 35mm, 55 mm. In this section, the obtained curves for the double magnet and the sum of two characteristics from Fig.1a were compared. This analysis is very important because it can present the possibility of separate consideration of interactions between each magnet and coil. If the superposition principle is confirmed then each interaction magnet-coil can be described by a separate mathematical relationship. The sum of two curves for a single magnet $U=f(x)$ (Fig.1a) must take into account their respective shift:

$$U_1=f(x_1) \quad where \quad x_1=x-L/2-l/2, \tag{4}$$

$$U_2=-f(x_2) \quad where \quad x_2=x+L/2+l/2, \tag{5}$$

where: U_1 is curve $U=f(x)$ taken from Fig.1a and shifted by distance $-L/2-l/2$ in coordinate x domain. Whereas U_2 is obtained from the same curve $U=f(x)$, it is inverted and shifted by distance

$L/2+l/2$ in coordinate x domain. This curve reversal is due to the orientation of magnet poles relative to the coil (NS or SN). Now, for the same motion direction and the same position relative coil, the considered separately single magnets will generate voltage with the opposite sign. Finally, the coil voltage from the superposition of both interactions can be written as:

$$U=U_1+U_2. \tag{6}$$

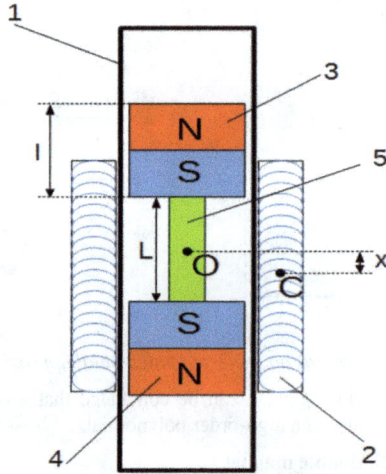

Fig.3. Scheme of experimental setup.

1 - tube, 2 - inductive coil, 3 - top magnet, 4 - bottom magnet, 5 - connector, C - coil center, O - connector center, L - connector length, l - magnet length, x- distance between coil center and double magnet center

a)

b)

c)

d)

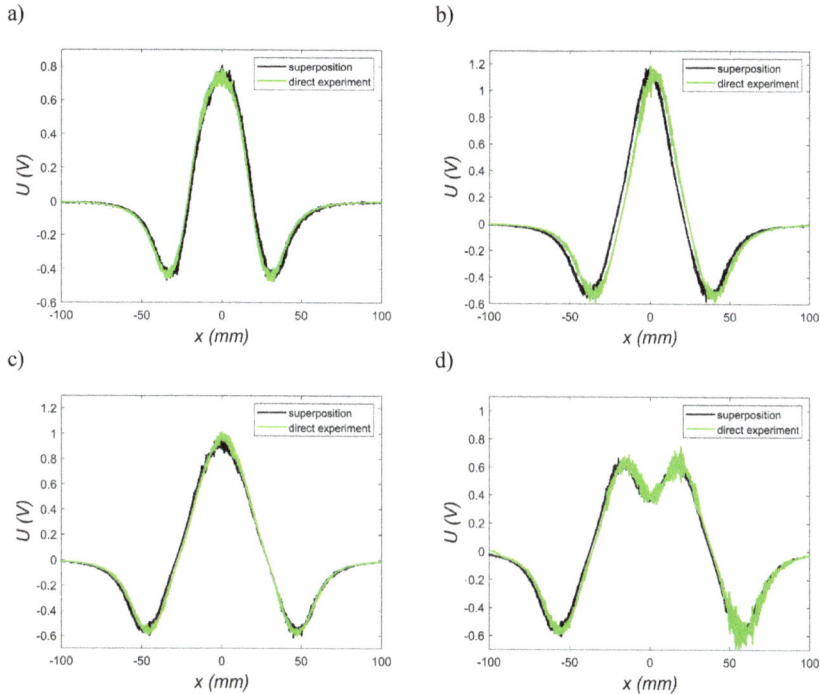

Fig. 4. Comparison of coil voltage from direct experiment for double magnet and superposition estimation (6).

a) L=0 mm, b) L=15 mm, c) L=35mm, d) L=55 mm.

The obtained results (Fig.4) show that the voltage measured for a system with a double magnet can be mapped using the superposition principle. Comparison of both cases (black and green series) have high compatibility. The maximum and minimum values are very similar. Only a very small shift of both signals can be observed. Based on this analysis, it can be concluded that a separate description of the magnet-coil interaction can be used in the modelling process. Finally, the mathematical description of coil voltage for a case with a double magnet can be written in the following form:

$$U=(\alpha_0+\alpha_1 x_1+\alpha_2 x_1^2+\alpha_3 x_1^3+...+\alpha_n x_1^n)v\pm(\alpha_0+\alpha_1 x_2+\alpha_2 x_2^2+\alpha_3 x_2^3+...+\alpha_n x_2^n)v, \qquad (7)$$

where v- double magnet speed and sign $\pm$ depends on magnet pole orientation: minus - repulsive magnets presented in this paper or plus - attracting magnets.

Summary
This paper is an important link between previous research [5] and the new concept. In the future, the system with two levitating magnets and one inductive coil will be considered. The presented results give some information, on how to create a new model. The curve of the electromechanical coupling coefficient is strongly nonlinear. It can be described by a polynomial function. However,

the polynomial order should be high ($n>24$). The high-order polynomial can generate some problems when the analytical solution will be searched. The polynomial can be replaced by a low-order trigonometric function [9], but this approach can generate even more problems during the analytical solution calculations.

In the mathematical description of the interaction between the top or bottom magnet and the coil can be investigated separately. It results from the analysis of the correctness of the superposition principle. This fact is important. If the superposition principle would not correct, then the model development of a system with two levitating magnets (case without connector) will be very difficult or impossible.

Acknowledgements
This work was financially supported under the project of the National Science Centre according to decision no. DEC-2019/35/B/ST8/01068.

References
[1] M. Mosch, G. Fischerauer, A comparison of methods to measure the coupling coefficient of electromechanical vibration energy harvesters, Micromachines, 10 (2019) 0823. https://doi.org/10.3390/mi10120826

[2] K. Kecik, A. Mitura, J. Warminski, S. Lenci, Foldover effect and energy output a nonlinear pseudo-maglev harvester, AIP Conf. Proc., 1922 (2018) 100009. https://doi.org/10.1063/1.5019094

[3] B. Sungryong, K. Pilkee, Load resistance optimization of bi-stable electromagnetic energy harvester based on harmonic balance, Sensors, 21 (2021) 1505. https://doi.org/10.3390/s21041505

[4] G.D. Pasquale, A. Soma, F. Fraccorollo, Comparison between piezoelectric and magnetic strategies for wearable energy harvesting, J. Phys.: Conf. Ser., 476 (2013) 012097. https://doi.org/10.1088/1742-6596/476/1/012097

[5] A. Mitura, K. Kecik, Modeling and energy recovery from a system with two pseudo-levitating magnets, Bull. Pol. Acad. Sci. Tech. Sci., 70 (2022) e121721.

[6] K. Kecik, A. Mitura, S. Lenci, J. Warminski, Energy harvesting from a magnetic levitation system, Int. J. Non-linear Mech., 94 (2017) 200-206. https://doi.org/10.1016/j.ijnonlinmec.2017.03.021

[7] https://www.mathworks.com/help/matlab/ref/polyfit.html

[8] S. Hoffman, Estimation of prediction error in regression air quality models, Energies, 14 (2021) 7387. https://doi.org/10.3390/en14217387

[9] K. Kecik, Modification of electromechanical coupling in electromagnetic harvester, Energies, 15 (2022) 4007. https://doi.org/10.3390/en15114007

Experimental Mechanics
Materials Research Proceedings 30 (2023) 61-67

Materials Research Forum LLC
https://doi.org/10.21741/9781644902578-9

Application of digital image correlation method to assess temporalis muscle activity during unilateral cyclic loading of the human masticatory system

Dominik Pachnicz[1,a*], Przemysław Stróżyk[2,b]

[1]Faculty of Mechanical Engineering, Łukasiewicza 5 str., 50-371 Wrocław, Wrocław University of Science and Technology, Poland

[2]Department of Mechanics, Materials and Biomedical Engineering, Smoluchowskiego 25 str., 50-370 Wrocław, Wrocław University of Science and Technology, Poland

[a]dominik.pachnicz@pwr.edu.pl, [b]przemyslaw.strozyk@pwr.edu.pl

Keywords: In Vivo Measurements, Optical Method, Masticatory Muscle Activity

Abstract. This paper presents an in vivo experimental study in which an optical digital image correlation system was used to assess the activity of the temporalis muscle. The muscle activity was analysed and assessed based on its displacements resulting from unilateral cyclic loading and unloading of a specimen placed on one side of the mandible between pairs of corresponding premolars and molars. Two sets of synchronised cameras (two per side) positioned on the working and non-working sides were used for the measurements. The results of the measurements were analysed individually for each part of the muscle, i.e. the anterior temporalis, the middle temporalis and the posterior temporalis and each side. The results indicate that the presented measurement method made it possible to determine temporalis muscle activity in vivo from displacement measurements. It also confirms the information on temporalis muscle function given by other researchers. In addition, the advantage of the presented method is that it offers significantly greater measurement capabilities (larger area of analysis) than other measurement methods, such as electromyography.

Introduction

Muscle activity or muscle force values are usually determined by electromyography (*EMG*) or numerical simulations. In the first method, the electrical potential of the muscle is measured, which is the determinant of its bioactivity. Force values are then calculated based on empirical equations relating a given potential value [1, 2] and the muscle's physiological active cross-section (*PCS*) [3, 4]. In numerical simulations, muscle forces can be determined by inverse kinematics and dynamics analysis [5, 6, 7, 8, 9].

During *EMG* measurements of the masticatory muscles, the temporalis and masseter muscles are most commonly studied because they are located externally from the buccal side, allowing easy access for electrode placement. In the case of the lateral pterygoid and medial pterygoid muscles, the application of *EMG* is very difficult because their location (hidden behind bony elements and soft tissues) requires the application of electrodes intraorally [10, 11].

Based on an analysis of the biomechanics and mechanics literature [12], a method based on 3D digital image correlation can also be used to measure surface muscle activity. The main advantage of the method is that it is possible to measure the activity of the entire muscle and not just a limited area lying around the electrode, as in *EMG* measurements. Disadvantages include the need to cover the surface of the test area with a suitable mask and illuminate it with monochromatic light.

The temporalis muscle was chosen for the study described in this article because: (1) it is easily accessible from the outside, (2) it is located in a zone with little adipose tissue, (3) anatomically,

Experimental Mechanics
Materials Research Proceedings 30 (2023) 61-67

Materials Research Forum LLC
https://doi.org/10.21741/9781644902578-9

it is the largest muscle in the head and also the largest muscle involved during mastication and (4) it is divided in a plane rather than by depth.

In anatomical terms, the muscle is divided into two areas according to the actions performed by its muscle fibres: the anterior part, characterised by a near-vertical fibre arrangement, is responsible for retracting the mandible, while the posterior part, characterised by a near-horizontal fibre arrangement, retracts the protruding mandible forwards. In the biomechanics of the masticatory system, the temporalis muscle is divided into two or three areas [13, 14, 15, 16]. Each area is most often modelled by a single vector (a component of the principal vector of the temporalis muscle).

This study aimed to identify areas of peak temporalis muscle activity in vivo during unilateral cyclic loading and unloading of the masticatory system. The analysis was performed for rhythmic changes in muscle displacement on the working and non-working sides using a 3D digital image correlation system.

Material and Methods

Determination of temporalis muscle displacement required the preparation of an experimental rig based on a non-contact optical digital image correlation system (*DIC* - Dantec Q400, Dantec Dynamics A/S, Skovlunde, Denmark), consisting of 2 synchronised camera sets to set up on the working (*W*) and balancing sides (*N*) - fig. 1. Initial testing was performed on an adult male subject after informed consent. The subject had full dentition with no known masticatory dysfunction. During the measurement, he sat comfortably in an upright position with his head resting against the headrest.

Fig. 1. The experimental setup used a digital image correlation system (DIC) to measure the temporalis muscle displacement during the masticatory system's unilateral cyclic loading and unloading.

Temporalis muscle activity was assessed based on its displacement during unilateral cyclic loading and unloading of a specimen (made of rubber-derived material) placed on one side of the mandible, between pairs of corresponding premolars (45-15) and molars (46-16) - fig. 2.

Experimental Mechanics
Materials Research Proceedings 30 (2023) 61-67

Materials Research Forum LLC
https://doi.org/10.21741/9781644902578-9

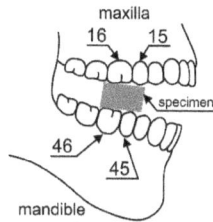

Fig. 2. Schematic showing the specimen placement between pairs of corresponding premolars (45-15) and molars (46-16).

Taking measurements with the *DIC* system required the temporalis muscle (i.e. the skin covering the temporalis muscle) to be covered with a unique pattern of black and white spots, which are used by the correlation algorithm as a source of information (Fig. 3). Prior to the final measurement, a reference photo (the so-called zero state) of the temporalis muscle was taken, the tension of which corresponded to the placement of the sample between the teeth.

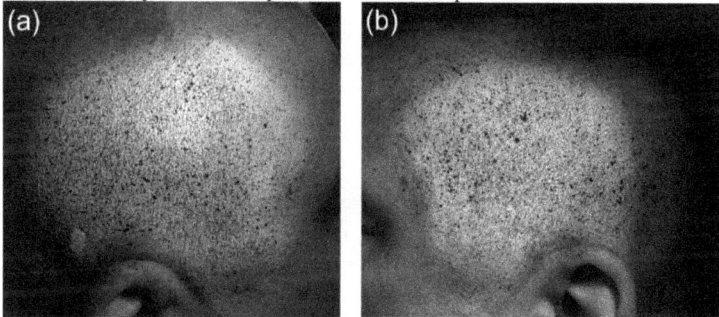

Fig. 3. Unique pattern of black and white spots; (a) working side and (b) non-working side.

During the test, the subject was asked to bite at a constant natural rate and a subjectively accepted force close to the maximum bite force. Measurements were taken three times for five cycles of loading and unloading. Images were taken at 4 Hz.

Results

The parameter analysed to determine the activity of the temporalis muscle was displacement perpendicular to the plane of the image, i.e. in the Z-axis direction (fig. 4). The results of the measurements were analysed from circular areas, defined on the surfaces of each of the three parts of the temporalis muscle, i.e. the anterior temporalis (AT), the middle temporalis (MT) and the posterior temporalis (PT) - fig. 4. The division of the muscle was carried out based on information reported in the literature related to the biomechanics of the temporalis muscle [17, 18]. The displacement values are the average of the results from a 10.0 [mm] diameter circle - fig. 4. The sites were selected based on the images obtained for the maximum displacement values for each side and part of the muscle, respectively, for the working side (p_{AWz}, p_{MWz}, p_{PWz}) and non-working side (p_{ANz}, p_{MNz}, p_{PNz}) - fig. 5. In addition, for each maximum displacement value, the standard deviations ($\pm SD$) were determined - Table 1.

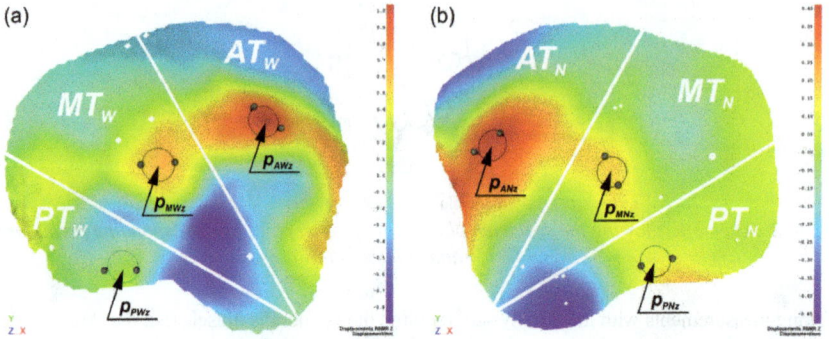

Fig. 4. Division of the temporalis muscle into three parts; (a) working side and (b) non-working side, and the areas (delimited by a circle) from which the mean values of temporalis muscle displacement were analysed.

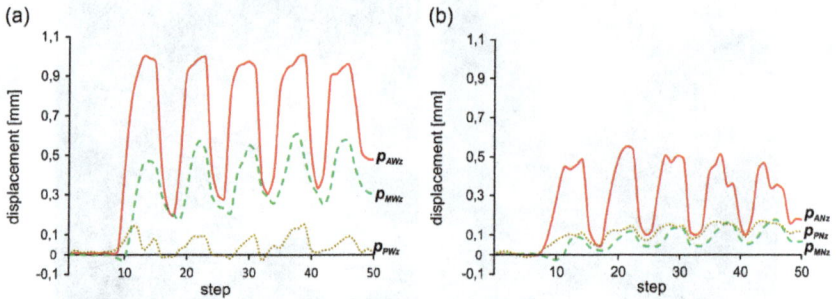

Fig. 5. Mean displacement values determined for the different parts of the temporalis muscle; (a) working side and (b) non-working side.

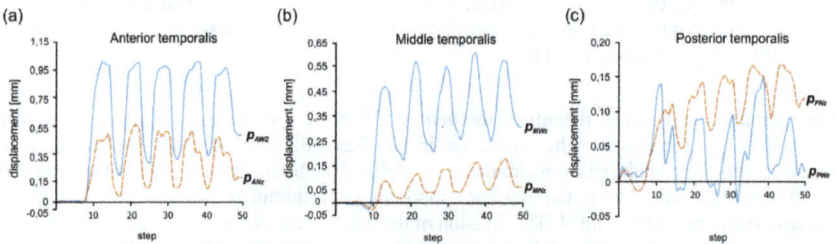

Fig. 6. Comparison of mean displacement values between the working side (p_{AWz}, p_{MWz}, p_{PWz}) and the non-working side (p_{ANz}, p_{MNz}, p_{PNz}) for the three parts of the temporalis muscle; (a) anterior temporalis, (b) middle temporalis and (c) posterior temporalis.

Experimental Mechanics
Materials Research Proceedings 30 (2023) 61-67

Materials Research Forum LLC
https://doi.org/10.21741/9781644902578-9

Table 1 Maximum, mean displacement values [mm] and standard deviation (±SD) for each part of the temporalis muscle, corresponding to the working side (p_{AWz}±SD, p_{MWz}±SD, p_{PWz}±SD) and non-working side (p_{ANz}±SD, p_{MNz}±SD, p_{PNz}±SD).

Step	Part of muscle		
	AT	*MT*	*PT*
	Working side		
	(p_{AWz}±SD)	(p_{MWz}±SD)	(p_{PWz}±SD)
1	0.99±0.010	0.47±0.001	0.08±0.010
2	0.99±0.010	0.57±0.002	0.08±0.003
3	0.97±0.010	0.55±0.002	0.09±0.001
4	0.95±0.006	0.59±0.004	0.12±0.013
5	0.94±0.004	0.57±0.008	0.07±0.013
	Non-working side		
	(p_{ANz}±SD)	(p_{MNz}±SD)	(p_{PNz}±SD)
1	0.46±0.030	0.09±0.009	0.11±0.030
2	0.55±0.060	0.11±0.006	0.13±0.030
3	0.50±0.003	0.13±0.007	0.14±0.020
4	0.50±0.002	0.14±0.004	0.16±0.030
5	0.47±0.003	0.17±0.007	0.16±0.020

Conclusions

This paper presents an experimental study in which the *DIC* system was used to assess the activity of the temporalis muscle during unilateral cyclic loading and unloading-of the masticatory system.

The most important result of the experimental studies carried out is the demonstration that the use of the *DIC* makes it possible to determine the activity of the entire temporalis muscle, and the results clearly show that the involvement of the muscle on the working side is different from that on the non-working side. This confirms the results presented in other work [14, 15, 19], in which similar results were obtained based on muscle forces. This means that there is a correlation between the results obtained from the *DIC* system and, for example, methods based on vector calculus.

The analysis of the mean displacement values shows that the anterior part on the working side is more involved than the other parts - similar correlations occur on the non-working side. Based on the maximum displacement values (Table 1), it was noted that on the working side, the temporalis muscle activity is significantly higher than on the non-working side for the *AT* and *MT* parts by 49% and 77%, respectively. For the *PT* part, the non-working side shows more activity than the working side by 59%. This means that the muscle is primarily responsible for lifting the mandible and pressing the teeth against the specimen on the working side. In contrast, on the non-working side, muscle activity is related to lifting (stabilising) and retracting the mandible. The cycles of muscle activity overlap in phase on both sides. This indicates that they work evenly and are involved in the masticatory function.

It can also be seen from the displacement analysis (Table 1 fig. 5) that on the working side *MT* and *PT* activity are lower than *AT* by 43% and 91%, respectively. On the non-working side, on the other hand, *MT* and *PT* are also smaller than *AT*, by 72% and 75%, respectively, with the difference that *PT* displacements are slightly larger than *MT*.

Experimental Mechanics   Materials Research Forum LLC
Materials Research Proceedings 30 (2023) 61-67   https://doi.org/10.21741/9781644902578-9

The study results indicate that the presented measurement method made it possible to determine temporalis muscle activity from displacement measurements. Furthermore, the method is an interesting complement to existing measurement methods, e.g. *EMG* [20, 21, 22] but offers a greater range of analysis than *EMG*. In addition, the software used by the *DIC* system offers the possibility to export the results to *FEM* (Finite Element Method) software and to compare the results between the real and virtual object, i.e. to perform validation.

The presented research needs to be further developed to expand the possibilities of using the *DIC* method to analyse masticatory system function. The number of subjects needs to be increased, allowing statistical processing of the results and more extensive inference. In addition, different load cases can be considered. Furthermore, finding correlations between the results obtained by both methods, i.e. vision and *EMG*, seems to be an important issue. The results of such considerations can be used, among other things, to determine the forces in the muscles and, after a more in-depth study of the subject, can also be useful in diagnosing the masticatory system.

References

[1] A.L.Hof, The relationship between electromyogram and muscle force. Sportverletz Sportsc. 11 (1997) 79–86, https://doi.org/10.1055/s-2007-993372

[2] G.J. Pruim, H.J. de Jongh, J.J. ten Bosch, Forces acting on the mandible during bilateral static bite at different bite force levels. J. Biomech. 13 (1980) 755–763, https://doi.org/10.1016/0021-9290(80)90237-7

[3] B. May, S. Saha, M. Saltzman, A three–dimensional mathematical model of temporomandibular joint loading. Clin. Biomech. 16 (2001) 489–495, https://doi.org/10.1016/S0268-0033(01)00037-7

[4] M. Radu, M. Marandici, T.L. Hottel, The effect of clenching on condylar position: a vector analysis model. J. Prosthet. Dent. 91 (2004) 171–179, https://doi.org/10.1016/j.prosdent.2003.10.011

[5] T.S. Buchanan, D.G. Lloyd, K. Manal, T.F. Besier, Neuromusculoskeletal modeling: estimation of muscle forces and joint moments and movements from measurements of neural command. J. Appl. Biomech. 20 (2004) 367–395, https://doi.org/10.1123/jab.20.4.367

[6] A.G. Hannam, Current computational modelling trends in craniomandibular biomechanics and their clinical implications. J. Oral. Rehabil. 38 (2011) 217–234, https://doi.org/10.1111/j.1365-2842.2010.02149.x

[7] P. Stróżyk, J. Bałchanowski, Effect of foodstuff on muscle forces during biting off. Acta Bioeng. Biomech. 18 (2016) 81–91, PMID: 27405536

[8] P. Stróżyk, J. Bałchanowski, Modelling of the forces acting on the human stomatognathic system during dynamic symmetric incisal biting of foodstuffs. J. Biomech. 79 (2018) 58–66, https://doi.org/10.1016/j.jbiomech.2018.07.046

[9] P. Stróżyk, J. Bałchanowski, Effect of foods on selected dynamic parameters of mandibular elewator muscles during symmetric incisal biting. J Biomech. 106 (2020) 109800, https://doi.org/10.1016/j.jbiomech.2020.109800

[10]P. Koole, F. Beenhakker, H. J. de Jongh, G. Boering, A standardized technique for the placement of electrodes in the two heads of the lateral pterygoid muscle. J. Craniomandib Pract. 8 (1990) 154–162, https://doi.org/10.1080/08869634.1990.11678309

Experimental Mechanics | Materials Research Forum LLC
Materials Research Proceedings 30 (2023) 61-67 | https://doi.org/10.21741/9781644902578-9

[11] G. M. Murray, T. Orfanos, J. Y. Chan, K. Wanigaratne, I. J. Klineberg, Electromyographic activity of the human lateral pterygoid muscle during contralateral and protrusive jaw, movements. Arch. Oral Biol. 44 (1999) 269–285, https://doi.org/10.1016/S0003-9969(98)00117-4

[12] Information on https://www.dantecdynamics.com/all-scientific-papers/

[13] M. Tuijt, J.H. Koolstra, F. Lobbezoo, M. Naeije, Differences in loading of the temporomandibular joint during opening and closing of the jaw. J Biomech. 43 (2010) 1048-54, DOI: 10.1016/j.jbiomech.2009.12.013, https://doi.org/10.1016/j.jbiomech.2009.12.013

[14] J.M. Reina, J.M. Garcia-Aznar, J. Dominguez, M. Doblaré, Numerical estimation of bone density and elastic constants distribution in a human mandible. J. Biomech. 40 (2007) 826–836, https://doi.org/10.1016/j.jbiomech.2006.03.007

[15] T.W.P. Korioth, D.P. Romilly, A.G. Hannam, Three-dimensional finite element stress analysis of the dentate human mandible. Am. J. Phys. Anthropol. 88 (1992) 69–96, https://doi.org/10.1002/ajpa.1330880107

[16] D. Luo, Q. Rong, Q.Chen, Finite-element design and optimization of a three-dimensional tetrahedral porous titanium scaffold for the reconstruction of mandibular defects. Med Eng Phys. 47 (2017) 176–183, https://doi.org/10.1016/j.medengphy.2017.06.015

[17] G.E.J. Langenbach, A.G. Hannam, The role of passive muscle tensions in a three-dimensional dynamic model of the human jaw. Arch. Oral Biol. 44 (1999) 557–573, https://doi.org/10.1016/S0003-9969(99)00034-5

[18] J.H. Koolstra, T.M.G.J. van Eijden, W.A. Weijs, M. Naeije, A three-dimensional mathematical model of the human masticatory system predicting maximum possible bite forces. J. Biomech. 21 (1988) 563–576, https://doi.org/10.1016/0021-9290(88)90219-9

[19] M. Pinheiro, J.L. Alves, The feasibility of a custom-made endoprosthesis in mandibular reconstruction: Implant design and finite element analysis. J Craniomaxillofac Surg. 43 (2015) 2116–2128, https://doi.org/10.1016/j.jcms.2015.10.004

[20] K. Aldana, R. Miralles, A. Fuentes, S. Valenzuela, M.J. Fresno, H. Santander, M.F. Gutiérrez, Anterior Temporalis and Suprahyoid EMG Activity During Jaw Clenching and Tooth Grinding, CRANIO®, 29 (2011) 261-269, https://doi.org/10.1179/crn.2011.039

[21] L. Lauriti, L.J. Motta, C.H.L. de Godoy, D.A. Biasotto-Gonzalez, F. Politti, R.A. Mesquita-Ferrari, K.P.S. Fernandes, S.K. Bussadori, Influence of temporomandibular disorder on temporal and masseter muscles and occlusal contacts in adolescents: an electromyographic study. BMC Musculoskelet. Disord.15 (2014) 123, doi: 10.1186/1471-2474-15-123

[22] A. Sabaneeff, L.D. Caldas, M.A.C. Garcia, M.C.G. Nojima, Proposal of surface electromyography signal acquisition protocols for masseter and temporalis muscles. Res Biomed Eng. 33 (2017) 324-330, https://doi.org/10.1590/2446-4740.03617

Experimental Mechanics

Materials Research Forum LLC

Materials Research Proceedings 30 (2023) 68-74

https://doi.org/10.21741/9781644902578-10

Growth of fatigue cracks in specimens welded under bending with torsion

Dariusz Rozumek[1,a] *, Janusz Lewandowski [2,b]

[1]Opole University of Technology, Mikolajczyka 5, 45-271 Opole, Poland

[2]Measurement and Automation Center S.A, Hagera 14A, 41-800 Zabrze, Poland

[a]d.rozumek@po.edu.pl, [b]janusz210@wp.pl

Keywords: Welding, Bending with Torsion, Fatigue Crack Growth, Hardness, Microstructure, Fillet Welds

Abstract. The paper presents the results of crack development in specimens welded from S355 steel under bending and torsional loading. Welded joints, as a method of inseparable joining of technical structures, are commonly used in many areas of human activity. The aim of the study was to analyse the influence of the shape of concave and convex fillet joints on the development of fatigue cracks. The experimental tests (fatigue tests) were performed with the use of the fatigue machine MZGS-100, at a constant amplitude of the moment $M_a = 9.20$ N·m, and the stress ratio R = - 1 with a load frequency of 28.4 Hz.

Introduction

Research on the durability of structures, materials and various types of connections (separable and inseparable) carried out in scientific units must answer the questions posed by designers and constructors, whether the applied solutions will be optimal. Optimal, fully understood areas of life, i.e. economy, security, rational management of material resources, technical equipment or human potential. The results of these tests should be helpful in determining the service life of each structure in which the failure occurrence is at the lowest possible level [1,2].

The method of inseparable joining of elements by welding is widely used. It includes products, structures and machines from all branches of industry. Therefore, the constant interest of scientists in this subject allows them to publish the results of their research and improve the quality and durability of the joined elements [3-5]. Materials used in industry and welded joints are not without material and welding defects. Therefore, they should be taken into account in the durability assessment. There are a number of publications discussing the influence of material, welding and geometric defects, as well as residual stresses [6,7]. The aim of the work is to present the results of fatigue crack development of T-shaped welded joints with fillet welds, made of steel S355 subjected to bending with torsion, taking into account the shape of the welds and the selected heat treatment.

Materials and Methods

The test specimens were made of structural steel grade S355, in the normalized state. This steel is widely used in industry, including the construction of ships, bridges, lifting devices, devices and construction of used in the mining industry, tanks and pipelines, etc. Table 1 shows the chemical composition of the material and Table 2 some mechanical properties. The starting material of the specimens was a drawn rod with a diameter of Ø 30 mm. Then, as a result of the performed machining and the TIG welding process, ready-made samples were obtained. T-welded joints, with fillet welds, were made in two variants of the weld face, i.e. concave and convex.

Experimental Mechanics
Materials Research Proceedings 30 (2023) 68-74

Materials Research Forum LLC
https://doi.org/10.21741/9781644902578-10

Table 1. Chemical composition (in wt %) of the S355 steel

C	Mn	Si	P	S	Cr	Ni	Cu	Fe
0.2	1.49	0.33	0.023	0.024	0.01	0.01	0.035	Balance

Table 2. Mechanical properties of the S355 steel

σ_y (MPa)	σ_u (MPa)	E (GPa)	v (-)	A$_s$ (%)
357	535	210	0.30	21

The experimental tests were carried out on specimens welded without heat treatment and on specimens after heat treatments. The heat treatment was performed by subjecting the specimens to annealing at the temperature of 630°C for 2 hours. Shapes and dimensions of the tested specimens are presented in Fig. 1.

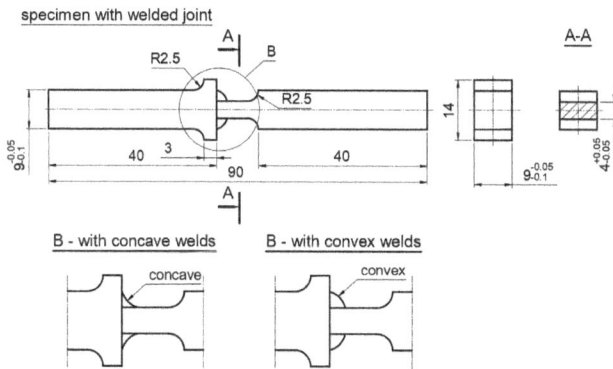

Fig. 1. Geometries of specimen with: (a) concave welds, (b) convex welds (dimensions in mm)

Metallographic tests were performed with the use of an optical microscope OLYMPUS IX70. Hardness measurements on the Vickers scale were carried out using a LECO MHT 200 hardness tester (LECO Corporation, St. Joseph, MO, USA), under a load of 100 g. The test to fatigue crack growth under cyclic bending with torsion were performed in the laboratory of the Department of Mechanics and Machine Design at Opole University of Technology on the fatigue test stand MZGS-100 [8,9] (Fig. 2). The loading method and the division into load components caused by bending and torsion are shown in Fig. 3. The tests were conducted under the amplitude of the total force moment control with the loading frequency of 28.4 Hz. The specimens restrained on one side were loaded with a constant amplitude of moments with the value M_a = 9.2 N·m and the load ratio R = - 1. The theoretical stress concentration factor, estimated with use of the model [10], in the solid specimen with concave weld under bending was K_t = 1.38 while it was 1.56 for the convex weld configuration.

Experimental Mechanics
Materials Research Proceedings 30 (2023) 68-74

Materials Research Forum LLC
https://doi.org/10.21741/9781644902578-10

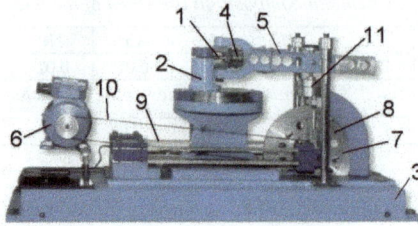

Fig. 2. MZGS-100 machine, where: 1 - specimen, 2 - rotational head with a holder, 3 - bed, 4 - holder, 5 - lever, 6 - motor, 7 - rotating disk, 8 - unbalanced mass, 9 - flat springs, 10 - driving belt, 11 - hydraulic connector

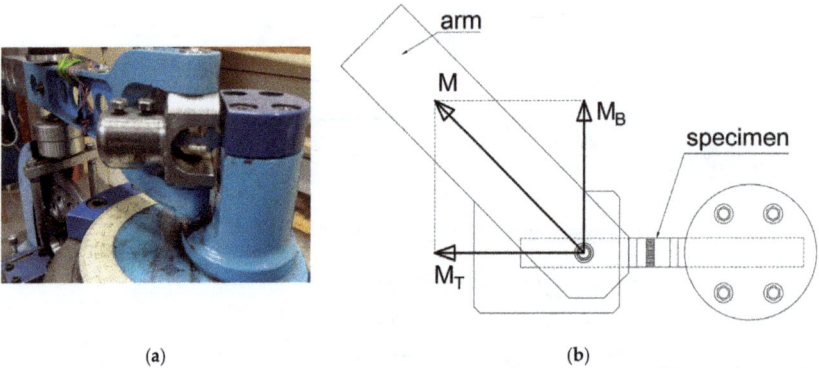

(a)   (b)

Fig. 3. The loading method and the division into load components caused by bending and torsion: (a) specimen clamped in MZGS-100 machine, (b) scheme of bending with torsion loading applied to the specimen

During the tests, the number of load cycles N was recorded. However the fatigue crack increments were measured with the micrometer located in the portable microscope with magnification of 20 times and accuracy up to 0.01 mm.

Results

As a result of the metallographic tests carried out in welded specimens, without heat treatment, a dendritic structure was observed in the areas of welds, and in the heat-affected zone (HAZ), a thick acicular structure of martensite and bainite was observed. On the other hand, in the specimens welded after heat treatment in the area of welds and HAZ, a coarse-grained structure of bainite and sorbite was observed (Fig. 4).

Experimental Mechanics
Materials Research Proceedings 30 (2023) 68-74

Materials Research Forum LLC
https://doi.org/10.21741/9781644902578-10

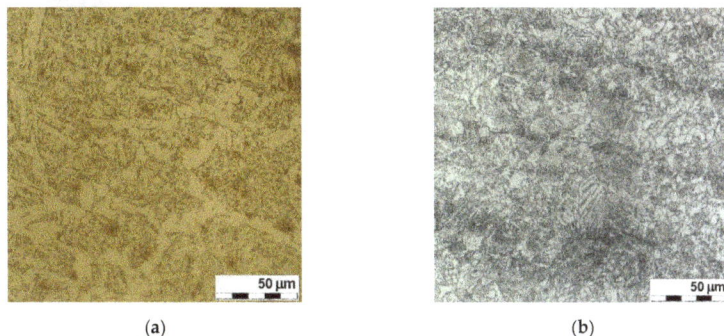

(a)   (b)

Fig. 4. The microstructure of welds in HAZ for a) without heat treatment, b) after heat treatment

The results of hardness measurements of welded specimens without heat treatment (HT) and after HT for the averaged results of specimens with concave and convex welds attained values:

- for specimens welded without HT, the hardness values changed significantly depending on the place of measurement. For the base material, the hardness remained the same (188–189 $HV_{0.1}$). While in the heat-affected zone (HAZ) large fluctuations in hardness were observed (194–248 $HV_{0.1}$). Then, moving to the weld metal, the measured values decreased and stabilized (230–220 $HV_{0.1}$).

- in the specimens subjected to relief annealing, the measured hardness and their variability were lower compared to the hardness of the specimens without HT. The smallest values were measured in the base material (about 135 $HV_{0.1}$), then in HAZ the hardness increased and ranged from 137 to 149 $HV_{0.1}$. However, the highest hardness values were measured in the weld material (about 150 $HV_{0.1}$).

Figure 5 presents the fatigue crack length versus number of cycles for proportional bending with torsion. In Fig. 5 can be observed that the longest fatigue life indicates specimens welded with concave welds without heat treatment.

Crack initiation (0.10 mm) was at 53000 cycles. The further development of the crack was rather quick and the specimens failed at 58000 cycles. The lowest fatigue life indicates the specimen with convex welds after normalizing (HT). Crack initiation (0.10 mm) was at 9000 cycles, and the failure of the specimens occurred at the number of cycles of 17,000.

Experimental Mechanics

Materials Research Forum LLC

Materials Research Proceedings 30 (2023) 68-74

https://doi.org/10.21741/9781644902578-10

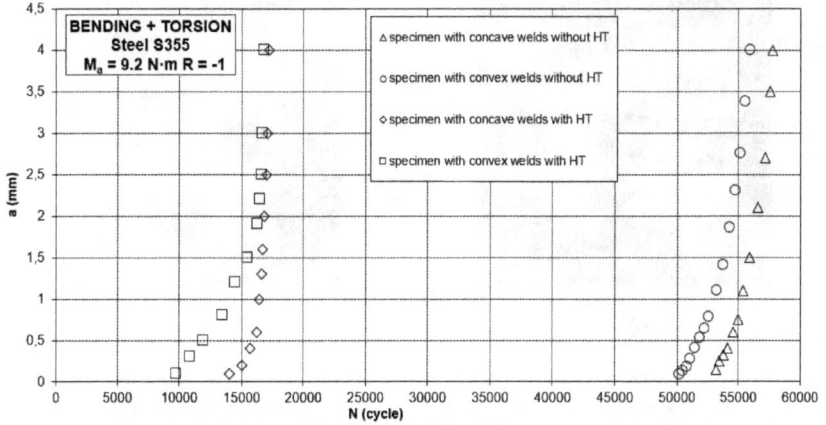

Fig. 5. Fatigue crack length vs. number of cycles under proportional bending with torsion

The differences in the fatigue life of the tested specimens welded with and without heat treatment are significant. In the case of specimens with concave welds, the decrease in fatigue life of relief annealed specimens was 69% compared to specimens without heat treatment. As in the case of specimens with concave welds, a 70% decrease in fatigue life of relief annealed specimens was observed for samples with convex welds compared to specimens without heat treatment. When comparing the durability of samples with concave and convex weld faces, for the same specimen (without heat treatment and after heat treatment), it can be seen that for specimens with concave welds, the durability was always higher compared to specimens with convex welds. The decrease in fatigue life of specimens with convex welds without heat treatment was 4.5% in comparison to specimens with concave welds. However the decrease in fatigue life of specimens with convex welds after heat treatment was 5.5% in comparison to specimens with concave welds. According to the authors, the significant drops in fatigue life in the specimens subjected to heat treatment were caused by structural changes taking place in the tested material. The higher durability of specimens with concave welds compared to specimens with convex welds is due to the occurrence of sharp notches in specimens with convex welds, which gave rise to cracks. The results of the fatigue life of the specimens without HT and with HT (annealing) obtained and presented in the publication are consistent with the results described in the work [11, 12], in which the authors examined the impact of heat treatment, i.e. hardening with tempering and annealing, on the fatigue life of steel.

Exemplary fatigue crack path is shown in Fig. 6 for a specimens without heat treatment with concave (a) and convex (b) welds and in Fig. 7 for a specimens with heat treatment. During laboratory tests, initiation and development of cracks occurred from one side of the specimen (from top or bottom) were observed, at the place of highest stress concentration, and after a certain period of propagation, the crack growth occurred also on the other side. The photos reported in Figs 6, 7 show crack paths whose shapes are typical of mixed modes I+III of fracture (bending with torsion). In all the considered cases the cracks were initiated perpendicular to the maximum normal stresses, in the fusion line, hence it can be assumed that the crack initiator was a geometric notch at the point of transition of the base material into the weld. After initiation, crack growth path develop along different planes depend on local stress state ahead of a crack tip strongly associated with weld geometry, microstructure and residual stresses after welding process.

Experimental Mechanics

Materials Research Forum LLC

Materials Research Proceedings 30 (2023) 68-74

https://doi.org/10.21741/9781644902578-10

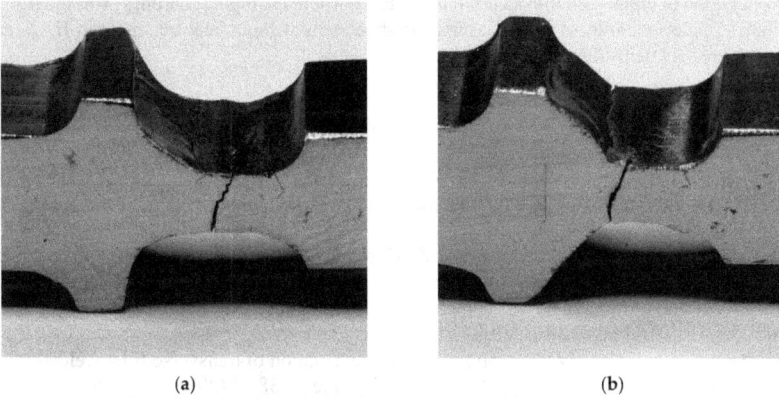

(a) (b)

Fig. 6. Examples of stages of crack development in a weld specimen without HT: (a) concave welds, b) convex welds

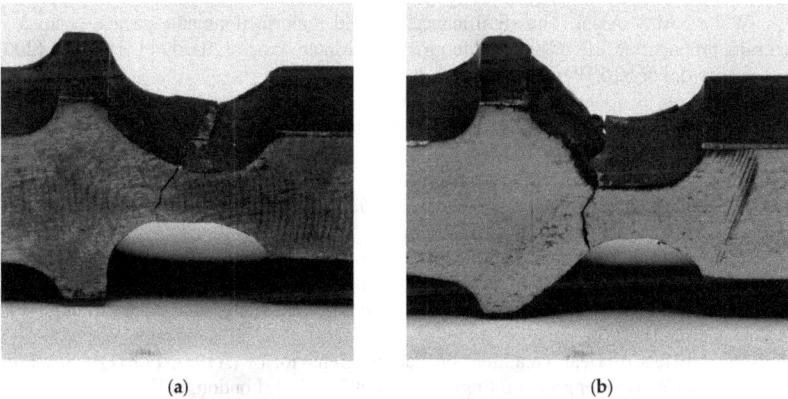

(a) (b)

Fig. 7. Selected stages of crack development in a weld specimen with HT: (a) concave welds, b) convex welds

Conclusions

The study presented the results concerning the fatigue crack growth in S355 steel specimens subjected to bending and torsion loading. The experimental outcomes allow to state the following conclusions:

- Fatigue life of the welded specimens without heat treatment were higher compared to the welded specimens with heat treatment, with slightly higher durability of specimens with concave welds.
- Initiation and fatigue crack growth in all test specimens started on one-side of the specimen, at the place of highest stress concentration, in the fusion line.
- The tested specimens, with heat treatment and without heat treatment, show different cracking courses.
- Cracking paths in tested specimens, with heat treatment and without heat treatment, show different courses.

Experimental Mechanics   Materials Research Forum LLC
Materials Research Proceedings 30 (2023) 68-74   https://doi.org/10.21741/9781644902578-10

- The propagation of cracks usually occurred in the HAZ where the highest hardness was measured.
- The highest material hardness was measured on specimens without heat treatment in HAZ, and the lowest in specimens after heat treatment.

References

[1] A. Carpinteri, C. Ronchei, D. Scorza, S. Vantadori, Fracture mechanics based approach to fatigue analysis of welded joints, Eng. Fail. Anal. 49 (2015) 67–78, https://doi.org/10.1016/j.engfailanal.2014.12.021.

[2] P.W. Marshall, Design of welded tubular connections. Basis and use of AWS code provisions, Elsevier, 1992.

[3] D. Rozumek, J. Lewandowski, G. Lesiuk, Z. Marciniak, J.A. Correia, W. Macek, The energy approach to fatigue crack growth of S355 steel welded specimens subjected to bending, Theoretical and Applied Fracture Mechanics, Vol 121 (2022), https://doi.org/10.1016/j.tafmec.2022.103470

[4] Z.-G. Xiao, T. Chen, X.-L. Zhao, Fatigue strength evaluation of transverse fillet welded joints subjected to bending loads, Int. J. Fatigue 38 (2012) 57–64, https://doi.org/10.1016/j.ijfatigue.2011.11.013.

[5] D. Rozumek, J. Lewandowski, G. Lesiuk, J. Correia, The influence of heat treatment on the behavior of fatigue crack growth in welded joints made of S355 under bending loading, Int. J. Fatigue 131 (2020), https://doi.org/10.1016/j. ijfatigue.2019.105328.

[6] M.A. Wahab, M.S. Alam, The significance of weld imperfections and surface peening on fatigue crack propagation life of butt-welded joints, J. Mater. Process. Technol. 153–154 (2004) 931–937, https://doi.org/10.1016/j. jmatprotec.2004.04.150

[7] Z. Jie, K. Wang, S. Liang, Residual stress influence on fatigue crack propagation of CFRP strengthened welded joints, Journal of Constructional Steel Research 196 (2022), https://doi.org/10.1016/j.jcsr.2022.107443

[8] D. Rozumek, S. Faszynka, Surface cracks growth in aluminum alloy AW-2017A-T4 under combined loadings. Eng. Fracture Mechanics 226 (2020) 106896, https://doi.org/10.1016/j.engfracmech.2020.106896.

[9] J. Lewandowski, D. Rozumek, Fatigue crack growth in welded S355 samples subjected to bending loading. Metals 11(9) (2021) 1394, https://doi.org/10.3390/met11091394.

[10] A. Thum, C. Petersen, O. Swenson, Verformung, Spannung und Kerbwirkung; VDI: Düesseldorf, Germany, 1960.

[11] M. Somer, Effect of Heat Treatment on Fatigue Behavior of (A193-51T-B7) Alloy Steel, Proceedings of the World Congress on Engineering 2007 Vol. II, London, U.K.

[12] Rafiq A. Siddiqui, Sayyad Z. Qamar, Tasneem Pervez, Sabah A. Abdul-Wahab, Effect of heat treatment and surface finish on fatigue fracture characteristics in 0.45% carbon steel, 10th International Research, Trends in the Development of machinery and Associated Technology, TMT 2006, Barcelona-Lloret de Mar, Spain.

Experimental Mechanics
Materials Research Proceedings 30 (2023) 75-82

Materials Research Forum LLC
https://doi.org/10.21741/9781644902578-11

Determination and experimental verification of the relation between stress amplitude and vibration amplitude in the VHCF regime

Piotr Swach[1,a*], Adam Lipski[1,b] and Michał Piotrowski[1,c]

[1] Bydgoszcz University of Science and Technology, Al. Kaliskiego 7, 85-796 Bydgoszcz, Poland

[a]piotr.swacha@pbs.edu.pl, [b]adam.lipski@pbs.edu.pl, [c]michal.piotrowski@pbs.edu.pl

Keywords: Very High Cycle Fatigue, Ultrasonic Testing, Strain Measurements, Stress Amplitude, Vibration Amplitude, Strain Gauge

Abstract. Fatigue tests conducted on ultrasonic machines are a relatively new testing method. The operation of test rig at a load frequency of 20 kHz and the numerical way of determining the relationship between the stress amplitude and an indirect method for determining the strain in the middle part of the sample causes researchers to feel uncertain about the stress value in the specimens. Purpose of this paper is to present experimental verification of the relationship between stress amplitude and vibration amplitude in the regime of Very High Cycle Fatigue, also the calibration procedure of the ultrasonic machine is demonstrated. The paper presents the methodology based on FEM that is used to determine the relationship between the stress amplitude in the smallest cross-section of the sample and the vibration amplitude for selected geometry.

Introduction

In the process of constructing machines and structures, it is necessary to take into consideration the issues related to material fatigue. The importance of this phenomenon had been already recognized in the early 19th century, while Wöhler's first experimental work appeared in the second half of that century. So far, this work, primarily due to research capabilities, has focused on low-cycle fatigue (LCF) and high-cycle fatigue (HCF) [1]. The least recognized is the range above 10^7 cycles, the so-called Very High Cycle Fatigue (VHCF) regime.

Until the late 1990s, the main problem with VHCF research was its time-consuming nature and the resulting high cost of research. This was due to the relatively low frequencies achieved by conventional testing machines. Therefore, attempts were made to develop devices that would allow work on much higher load frequencies. In 1950s, Mason [2,3] developed a piezoelectric transducer that converts a 20 kHz electrical signal into a mechanical wave of the same frequency. This solution was tried to be used in fatigue tests but controlling the device at such high frequencies was a significant obstacle. Only the appearance of efficient computers at the end of the 20th century allowed the development of an efficient research system that reduced the time of fatigue tests by about 1000 times in comparison to conventional machines. The development of a computer control system for ultrasonic fatigue testing machines by Bathias, Wu, and Ni [4] can be considered as the moment when commercially viable VHCF research has begun.

In constructions such as car and marine engines, turbine components, high-speed railway drive elements [5] and helicopter speed reducers [6] fatigue life is greater than 10^7 cycles. For testing of this type of elements, time reduction of a single test and the relatively low energy requirements make ultrasonic fatigue testing the only economically reasonable testing method for VHCF. An ultrasonic fatigue testing system uses the phenomenon of resonance to generate stresses in the specimen. The relationship between the stress amplitude in center section of specimen for ultrasonic testing and the vibration amplitude is linear, and its determination, for each specimen design is necessary to define the stress during the test controlled by the vibration amplitude.

The aim of this paper is to present methodology of determining, using FEM, the relationship between the stress amplitude in the smallest specimen cross-section and the vibration amplitude

Experimental Mechanics
Materials Research Proceedings 30 (2023) 75-82

Materials Research Forum LLC
https://doi.org/10.21741/9781644902578-11

for selected materials and geometries. The methodology of calibration of the ultrasonic fatigue machine is presented. The vibration amplitude values obtained during the calibration will allow to determine the machine's operating range, while the values obtained during the experimental verification will allow to determine the correctness of the relationship between the stress amplitude and the vibration amplitude.

Test station and method

In an ultrasonic machine (Fig. 1), a 20 kHz electrical signal produced by a generator is converted into a mechanical wave of the same frequency in a piezoceramic transducer, the amplitude of the vibration of this wave is amplified through a booster and high gain sonotrode forces the specimen to vibrate. Specimen should be designed so that its natural frequency is 20 kHz. During real-time testing, a displacement sensor measures the vibration amplitude of the specimen.

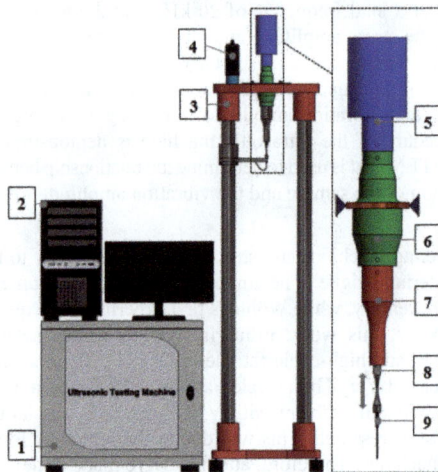

Fig. 1. Ultrasonic testing machine: 1 –generator 2.2 kW, 2 – PC unit, 3 – frame, 4 – air cooler, 5 – piezo-electric transducer, 6 – booster, 7 – sonotrode, 8 – specimen, 9 – displacement sensor

Calibration

To conduct tests on ultrasonic testing machine, it is necessary to calibrate it. Calibration allows you to determine the relation of the change in the supply voltage of the piezoceramic transducer to the amplitude of the displacement. The simplest way to perform calibration is to make a calibration rod in the form of a straight rod. To design the rod, analytical formulas can be used to define its length l:

$$l = \frac{1}{2f}\sqrt{\frac{E}{\rho}}, \qquad (1)$$

where: f – frequency, E – Young modulus, ρ – density.

Analyzing the formula above, it can be seen that with increasing sample vibration frequency, the length of the sample decreases. The value of the loading frequency of 20 kHz used in most ultrasonic machines is not accidental. At 20 kHz, the specimen length dimension for metallic materials does not cause fabrication problems.

Using the analytical formulas for a simple cylindrical bar, it is possible to determine the values of displacements u, strains ε and stresses σ:

$$c = \sqrt{\frac{E}{\rho}}, \tag{2}$$

$$k = \frac{\pi}{l}, \tag{3}$$

$$\omega = \frac{\pi c}{l}, \tag{4}$$

$$u(x) = A_0 \cos(kx), \tag{5}$$

$$\varepsilon(x) = -kA_0 \sin(kx), \tag{6}$$

$$\sigma(x) = -EkA_0 \sin(kx), \tag{7}$$

where: A_0 – maximum displacement amplitude.

Fig. 2. shows examples of displacement and stress distributions along the length of a straight bar of length l.

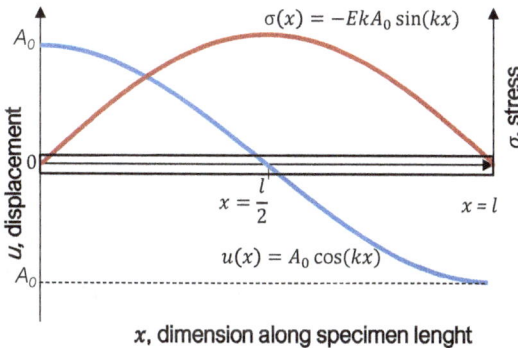

Fig. 2. Displacement and stress variation in a cylindrical bar

As it is shown above, the maximum stress of the specimen are obtained in its central section. Due to the fact that during the vibrations of the specimen, there is in a second mode of natural vibrations, then at both its ends there is an equal displacement amplitude. Using this information calibration can be made by fixing the rod in the system and using a high-resolution displacement sensor to determine the relation of the supply voltage on the amplitude of the displacement at the end of the rod. During calibration, it is important that the range of achievable vibration amplitude does not go beyond the elastic range of the material [7]. Fig. 3 shows the relationship of the supply voltage to the vibration amplitude. From the relation we can also read the range of vibrations in which the ultrasonic system is able to operate. Knowledge of the minimum and maximum amplitude of vibration, is essential to properly design a specimen for fatigue testing.

Experimental Mechanics

Materials Research Proceedings 30 (2023) 75-82

Materials Research Forum LLC

https://doi.org/10.21741/9781644902578-11

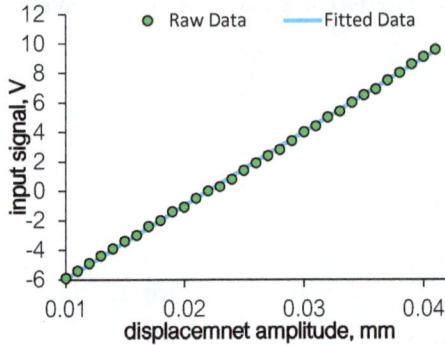

Fig. 3. Calibration curve of the ultrasonic machine presenting relation between the input voltage signal (V) to displacement amplitude (mm)

Properly performed calibration of the device enables testing. The procedure for conducting ultrasonic tests can be divided into 3 basic stages:
a) preliminary tests aimed at identification of basic properties of the tested material,
b) specimen geometry analysis using the finite element method (FEM),
c) main ultrasonic fatigue test.

The procedure is shown in Fig. 4.

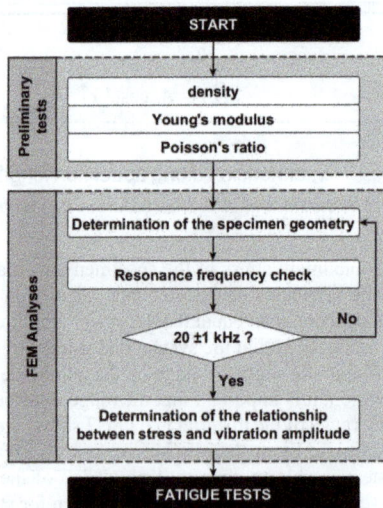

Fig. 4. Flow diagram of the test procedure

Preliminary tests include the determination of three basic strength parameters: Young's modulus (E), Poisson's ratio (v), density (ρ).

Experimental Mechanics   Materials Research Forum LLC
Materials Research Proceedings 30 (2023) 75-82   https://doi.org/10.21741/9781644902578-11

In the case of used specimen with a variable cross-section, the value of the stress amplitude is difficult to determine analytically. Therefore, in practice, the finite element method (FEM) is used for this purpose. As part of the second stage, a modal analysis is carried out using the FEM software, including the verification of the resonant frequency for the specimen geometry and the determination of the coefficient ks, which is the relationship between the vibration amplitude A_0 and the stress amplitude σ_a generated in the sample under the influence of vibration:

$$ks = \frac{\sigma_a}{A_0}. \tag{7}$$

The coefficient ks determined numerically is implemented into the test system, which after calibration calculates what voltage should be applied to the piezoceramic transducer to obtain the required vibration amplitude. In the last step of the procedure, proper fatigue tests are carried out.

When making specimens for ultrasonic testing, high quality machining is extremely important, as even a small error in geometry or too much roughness can disrupt its vibration, change its natural frequency and affect the test result. It is also important to verify that the ks factor has been determined correctly. Therefore, experimental verification of numerically determined stresses is necessary before fatigue testing.

Experimental measurement

Measurement system. On the test specimens, the strain gauge was glued at the location of the highest stresses, i.e. in the middle part of the reduced section with the smallest diameter. HBM 1-LY41-1.5/120 strain gauges with a gauge factor $k = 2$ were used in the tests.

Unfortunately, most strain gauge amplifiers have a sampling frequency value up to 20 kHz and a 20 kHz low-pass filter. Therefore, it was decided to build its own strain gauge amplifier to allow amplification in the bandwidth greater than 20 kHz. The correct operation of the circuit was checked using the function generator, which was given an electrical signal with a sinusoidal waveform, frequency 20 kHz and amplitude 0.5 V. The recorded output signal was compared with the result of the simulation of the circuit from the LTspice XVII program. The individual elements of the measurement system and the process of verifying the correct operation of the circuit are shown in Figure 5.

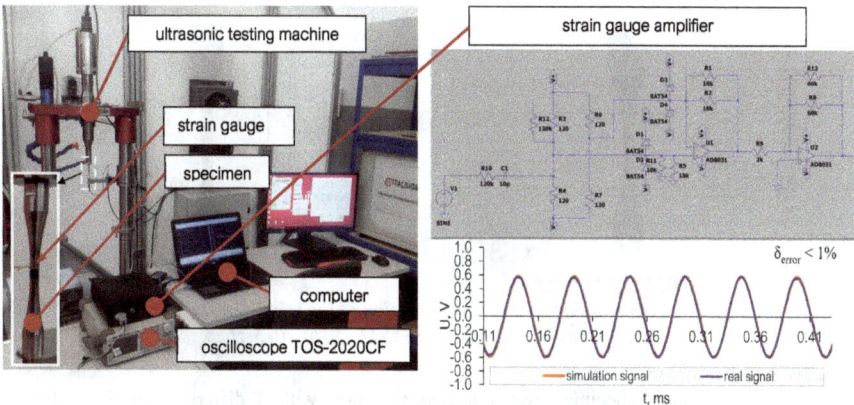

Fig. 5. Measurement system and result of circuit verification

To determine the stress amplitude σ_{atens}, the presented formula was used:

$$\sigma_{atens} = \frac{U_{pp} \cdot C_a \cdot E}{2 \cdot U_s},$$

(8)

where: U_{pp} – peak to peak voltage, U_s – supply voltage, C_a – calibration constant.

Testing materials. Specimens made of two different materials and geometries were used. The material properties are presented at Table 1.

Table 1. Basic material properties of the specimens

Material	Young's modulus, E	Poisson's ratio, v	Density, ρ
	GPa	-	kg/m^3
structural steel, S355J2+N	197.27	0.27	7820
aluminum alloy, 7075 T6	72.00	0.32	2800

The geometry of the specimens (Fig. 6) and the stress values for vibration amplitude $A_0 = 15$ μm were determined in ABAQUS software (Fig. 7). Values of the ks coefficient were determined for each of the specimen using the formula (7).

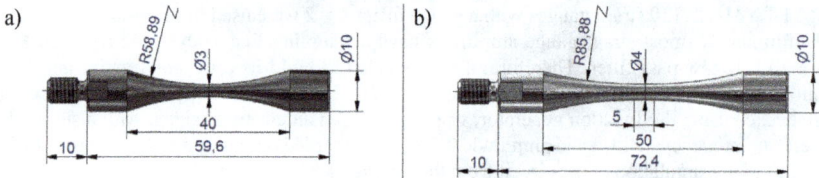

Fig. 6. Main dimensions of the specimens in mm: a) S355J2+N, b) 7075 T6

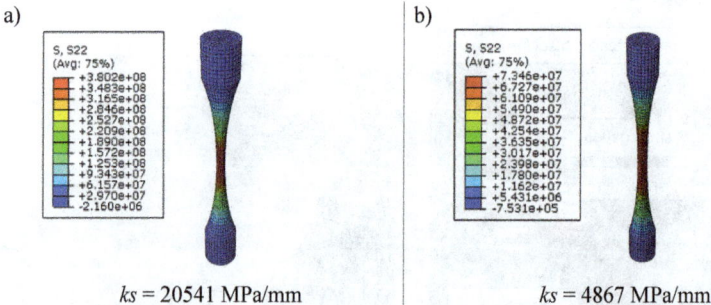

$ks = 20541$ MPa/mm $\qquad$ $ks = 4867$ MPa/mm

Fig. 7. Stress distribution in the axial direction of the specimens and the values of the ks coefficient: a) S355J2+N, b) 7075 T6

Results and discussion

During the tests, each specimen was put into vibration with 3 different amplitude levels. The recorded waveforms of strain changes ε at time t for S3555J2+N steel and for 7075 T6 aluminum alloy are shown in Figure 8.

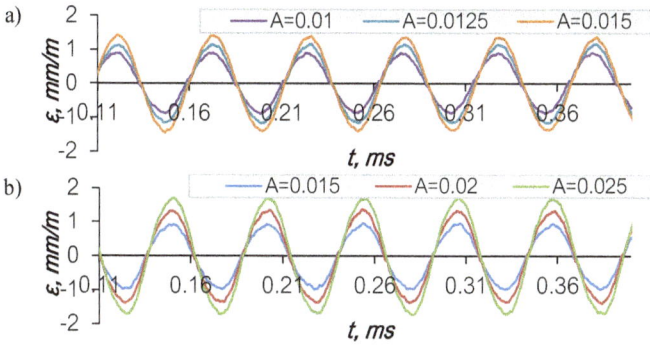

*Fig. 8. Changes in strain values for individual vibration amplitudes
of steel (a) and aluminum alloy (b) specimen*

The specimens were loaded for a short time to minimize the risk of strain gauge damage. Stresses for given vibration amplitudes were calculated using formula (7) and (8). The results and percent error are presented in Table 2. The results in the form of a diagram are shown on the Fig. 9.

Table 2. Basic material properties of the specimens

Parameters	S355J2+N			7075 T6		
	I	II	III	I	II	III
A_0, mm	0.015	0.02	0.025	0.01	0.0125	0.015
σ, MPa	73.01	97.34	121.68	205.41	256.76	308.12
σ_{tens}, MPa	72.75	96.59	119.84	197.29	248.45	303.20
$\delta = \frac{\sigma - \sigma_{tens}}{\sigma} \cdot 100, \%$	0.35	0.77	1.51	3.95	3.24	1.59

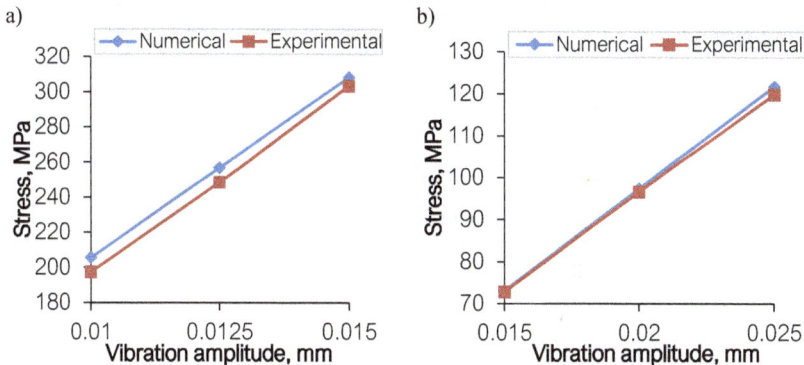

*Fig. 9. Vibration amplitude as a function of stress determined in the experiment and numerical
calculations for: a) S355J2+N, b) 7075 T6*

Experimental Mechanics   Materials Research Forum LLC
Materials Research Proceedings 30 (2023) 75-82   https://doi.org/10.21741/9781644902578-11

The results of the stress from the experimental measurements for steel have a small percentage mistake. The designed steel specimen has an hourglass shape, and the maximum stress occur in the center of the specimen, so probably the apparent difference in stress value is due to averaging it over the base length of the strain gauge. To verify this, the change in strain along the length of the test specimens for $A_0 = 15$ μm was determined using FEM software (Fig. 10). For aluminum, the experimental results and numerical results were very similar. The specimen in the central part has a cylindrical section of 5 mm, which causes a uniform distribution of strain along this length, so that averaging over the length of the strain gauge does not have such an impact on the result.

a)   b)

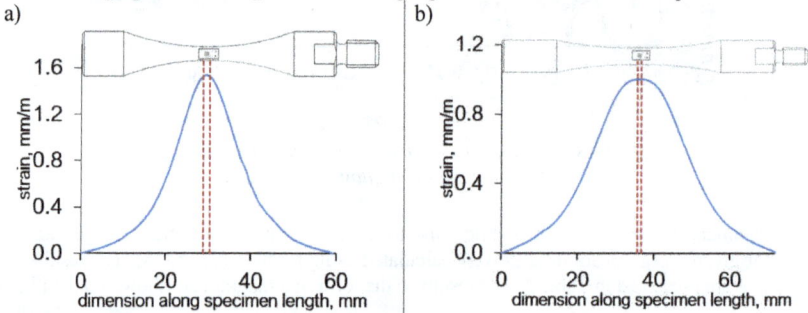

Fig. 10. Dependence of strain changes along the axis of the specimen for: a) S355J2+N, b) 7075 T6

Conclusion

The presented work is a part of the verification and validation of the research procedure implemented in the Laboratory for Research on Materials and Structures in the Bydgoszcz University of Science and Technology. To sum up the above work:

1. The conducted tests confirmed that the specimens were prepared correctly and the determined *ks* relations are correct.
2. Proper calibration of the ultrasonic machine is necessary to carry out fatigue tests.
3. The presented strain gauge system allows measurement of strain for frequencies of 20 kHz.
4. Measuring the strain of hourglass-shaped specimens using the strain gauge method, the accuracy of the result decreases due to their averaging over the length of the strain gauge base. It should also be remembered that for strain gauge measurements it is necessary to apply the strain gauge precisely in the proper location.

References

[1] Kocańda S., Szala J., Basic of fatigue calculation. Polish Scientific Publishers PWN, Warsaw 1997.

[2] Mason W.P., Piezoelectric crystals and their application to ultrasonics. Van Nostrand, New York 1950.

[3] Mason W. P., Internal Friction and Fatigue in Metals at Large Strain Amplitudes, J Acoust Soc Am, Vol. 28, No. 6, p. 1207, 1956. https://doi.org/10.1121/1.1908595

[4] Bathias C., Ni J., Wu T., An Automatic Ultrasonic Fatigue Testing System for Studying Low Crack Growth at Room and High Temperatures. STP 13973S, ASTM International, 1994.

[5] Bathias C., Paris C., Gigacycle fatigue in mechanical practice. New York, Marcel Dekker, 2004. https://doi.org/10.1201/9780203020609

[6] Shaniavski A. A., Skvortsov G. V., Crack growth in the gigacycle fatigue regime for helicopter gears, Fatigue Fract Eng Mater Struct, Vol. 22, No. 7, pp. 609–619, 1999. https://doi.org/10.1046/j.1460-2695.1999.00188.x

[7] Swacha P., Lipski A., Cracking of S355J2+N Steel in the High-Cycle and Very-High-Cycle Fatigue Regimes, Int J Fatigue, p. 107388, 2023. https://doi.org/10.1016/j.ijfatigue.2022.107388

Experimental Mechanics
Materials Research Proceedings 30 (2023) 83-90

Materials Research Forum LLC
https://doi.org/10.21741/9781644902578-12

Development of low-cost high-frequency data acquisition system for energy harvesting applications

Rafał Mech[1,a*], Oleksandr Ivanov[2,3,b], Przemysław Wiewiórski[1,c] and Bianka Kowalska[1,d]

[1]Wrocław University of Science and Technology, Wybrzeże Wyspiańskiego 27, 50-370 Wrocław, Poland

[2]Institute of Low Temperature and Structural Research PAS, Okólna 2, 50-422 Wrocław, Poland

[3]B.Verkin Institute for Low Temperature Physics and Engineering of NAS of Ukraine 47 Nauki Ave., 61103 Kharkiv, Ukraine

[a]rafal.mech@pwr.edu.pl, [b]o.ivanov@intibs.pl, [c]przemyslaw.wiewiorski@pwr.edu.pl, [d]bianka.koww@gmail.com

Keywords: Smart Materials, Magnetostriction, Terfenol-D, Wireless Sensors, Ultrasonic System

Abstract. The presented work describes a method in which, using a dedicated system, it is possible to simultaneously transfer energy and data between two devices. The proposed solution allows for supplying power to the sensor with simultaneous data transmission. The power transmission mechanism is based on the excitation of the structure with a wave, which is converted into electricity by a harvester device. Data transmission is carried out using the Double Frequency F2F procedure, which is a type of frequency modulation.

Introduction

In the last few decades, it has been possible to reduce the power needed to supply electronic devices to just a few dozen milliwatts [1]. At these power levels, traditional batteries are limited to short-term operation, mainly due to dimensional limitations. In addition, in the event of prolonged use, the batteries need to be replaced or recharged, and at the same time, they degrade.

Energy Harvesting (EH), originally known as energy harvesting or energy scavenging, is a set of techniques that provide electricity by converting energy from various sources such as mechanical, thermal, solar, and electromagnetic energy and salinity gradients, etc., e.g., [2]. In general, the main goal is to use sources commonly available in the environment, which in most cases are undesirable and suppressed (e.g. or as a result of electric current flow and engine cooling, etc.). Currently, it is said that EH can be a useful source of "cheap or no-cost" (excluding installation costs) power to low-power electrical devices [3-7].

One source of energy wastage is structural vibration. Vibrations in engineering structures such as buildings or bridges have low amplitudes and frequencies (0.1 g and 0.1 Hz); at the same time, various small electrical appliances, such as ovens, microwave ovens, and others, have higher amplitudes and frequencies (0.5 g and about 150 Hz, respectively) [8]. The above conditions inspired the development and description of many types of combined harvesters in the literature.

Two types of harvester devices can be distinguished, i.e. passive materials and active materials. In the case of active harvesters, most devices are based on magnetostrictive or piezoelectric materials (PZT) [9, 10]. It should be noted here that piezoelectric harvesters are capacitive energy sources; therefore they have a high output impedance. This means that appropriate power management circuits must be used to power electrical devices. On the other hand, there are magnetostrictive harvesters that are inductive. Thanks to this, they can provide low impedance at frequencies characteristic of most common sources of vibration.

Experimental Mechanics
Materials Research Proceedings 30 (2023) 83-90

Materials Research Forum LLC
https://doi.org/10.21741/9781644902578-12

Of the passive vibrational energy harvesters, the magnetostrictive devices deliver a higher energy density. Moreover, a comparison of magnetostrictive devices with devices based on piezoelectric material showed that both can generate similar levels of energy output; however, there is no need for additional special power management circuitry for solutions based on magnetostrictive material. Among magnetostrictive devices, the two most common types are the axial type and the bending type, based on the state of stress in the material. Axial devices are usually mounted in places where there is a high excitation force [11–19]. Thanks to such high loads, they can generate relatively high power densities, even up to 10 W/cm^3 [14]. In contrast to axle harvesters, the bending device for obtaining vibration energy can be mounted on any source of vibration [20–24]. It should be noted that in the case of magnetostrictive harvesters, their efficiency is variable and depends on many factors, such as load, operating frequency, mounting method, or the material from which the device is built.

In the literature on the subject, it can be seen that the research focused mainly on piezoelectric transducers [4–6,25,26]. It turns out, however, that in some cases a better solution is the use of magnetostrictive harvesters [27]. Taking into account the research conducted so far in this area, the main goals of this research are:

- development of a system enabling the transmission of energy and information through a solid body in the case of ultrasonic frequencies (the system should operate at frequencies above 20 kHz, ie inaudible to most people);
- use of intelligent materials (piezoelectric and magnetostrictive);

The paper presents a data acquisition system using an axle harvester. Such a harvester was selected after analyzing the already developed combines that can be found in the literature. In addition, such a solution was also associated with the expected amount of energy that could be generated using this type of device.

Magnetostrictive harvester

In the case of magnetostrictive devices, the most important component is the magnetostrictive core. Such a core may consist of one or more elements, depending on the size, length, and purpose of the device. The core material is also important. Such material must be characterized by gigantic magnetostriction (GMM - Giant Magnetostrictive Materials, e.g. Terfenol-D, nano-cobalt ferrite). An additional element is a system that allows you to adjust the initial magnetization of the material, which is usually properly selected neodymium magnets. The number of elements that make up the core of the device has a significant impact on the frequency of operation. The smaller the number of elements, the higher the operating frequency can be. To optimize the design of the actuator, magnetostrictive material and neodymium magnets are used alternately.

The actuators/harvesters presented in the article have relatively large dimensions: the diameter of the device was 44 mm and its height was 47 mm. The geometry was forced mainly by the need to apply the appropriate pressure of the cylinder to the tested structure. Inside the housing, there was a coil with resistance Rcoil = 5.5 Ω. The devices operated over a wide frequency range from 10 Hz to 30 kHz to find the resonant frequency of the system within which the system achieved the highest voltage values. In addition, both the harvester and the actuator were pre-tensioned with a force of 400 N. This value was determined based on experimental tests, during which the magnetomechanical response of the system was determined depending on the applied load. Successive constructions and modifications allow for more and more power; therefore, these devices began to be used as a source of electricity (Fig. 1).

Experimental Mechanics
Materials Research Proceedings 30 (2023) 83-90

Materials Research Forum LLC
https://doi.org/10.21741/9781644902578-12

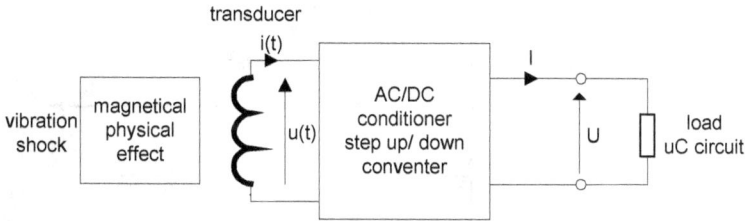

Figure 1. Harvester scheme.

Power and data transmission

In previous studies by the authors, it turned out that magnetic materials of the SMART type can be effectively used for the wireless transmission of energy and information [28]. The obtained results also indicate the high efficiency of this method. The system developed by the authors, which allowed the simultaneous transmission of data and power, is called SURPS (SMART Ultrasonic Resonant Power System). This system provides transmission through various solids as well as through liquids. An additional advantage of the system is the possibility of using various transmitter-receiver configurations [28]. The results presented in this work are a continuation of the work on the use of the magnetomechanical effect in the case of energy acquisition, described in more detail in [28].

The transmission was using an actuator that transfers the mechanical energy in the form of a pure sinusoidal ultrasonic wave, and then this wave was picked up by a harvester which converted this wave into an electric current through the magneto- or electrostrictive material contained therein. This way of transmitting energy also made it possible to transmit information. Moreover, it was possible to transmit energy through various materials, and the choice of material depended mainly on the distance over which the energy had to be transmitted.

For the transmission of information, the F2F procedure was used, which is a type of frequency modulation. The modulation worked in such a way that the data transmission frequency was an order lower than the structure's resonant frequency. Figure 2 shows schematically how the data was transmitted by the magnetostrictive actuator (AT) and how the signal was received by the magnetostrictive harvester.

System Structure

The newly developed system has been designed to work with various actuator-harvester configurations. One such configuration is a system where harvesters are connected in series and placed between two parallel beams. Such a system is characterized by a resonant frequency above 20 kHz. The test stand, consisting of two steel rails with magnetostrictive transducers placed between them, is shown in Fig. 3. This system enabled simultaneous powering of the microprocessor on the side of the power harvester and data transfer in both directions.

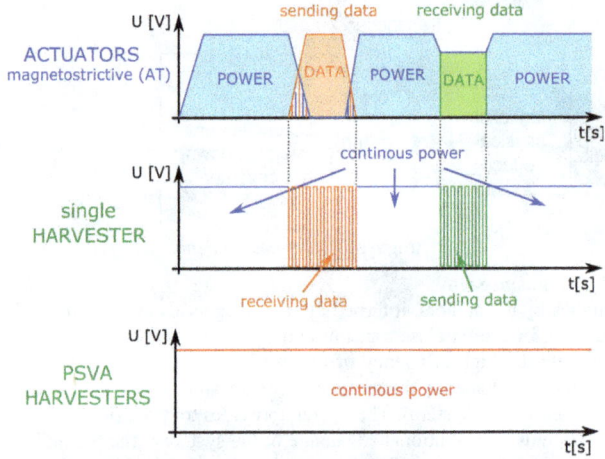

Figure 2. Scheme of data transmission and receiving information.

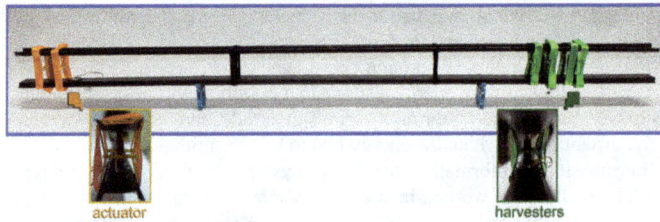

Figure 3. Actuator–harvester magnetostrictive system based on two beams.

Based on the above-described solution, supplemented with the current state of knowledge in the field of ultrasonic techniques and wireless power transmission, an original and innovative transmitting–receiving system was developed. The platform consists of a real-time module (STM32F411VE board) and a computer running Windows OS. A FTDI-bridge FT2232H was used to transfer data between the STM32-board and a personal computer. Figure 4 shows, the scheme of the system. Compared to Arduino-like solutions, this solution allows to reduce of the reaction time of the STM32-board to external events and provides a good data transfer rate (up to 8 Mbyte/Sec at FT245-Style Asynchronous FIFO mode). For software development, the non-commercial versions of IDE were used: Keil MDK 5.36 Community and Microsoft Visual Studio 2022 Community. The frequencies of the ultrasonic actuators were set in the standard way using an AD9850 DDS-generator. For each DDS frequency, 2048 16-bit words of ADC1-channel and 2048 16-bit words of ADC2-channel were recorded. To estimate the phase shifts, the IEEE standard three-parameter algorithm was used. Despite the non-optimal configuration of the ADC of STM32F411 (sequential sample and hold for each channel), a quantization frequency of 200 kHz was obtained for the two-channel measurement mode and transmission of a data packet through FTDI-bridge to the memory of a personal computer. This result was obtained for the layout on the contact breadboard and without software optimization. Optimal installation of modules and optimization of the program will increase the quantization frequency up to 500 kHz for the two-channel ADC mode and allow to make FFT analysis of the properties of different sensors in the frequency domain.

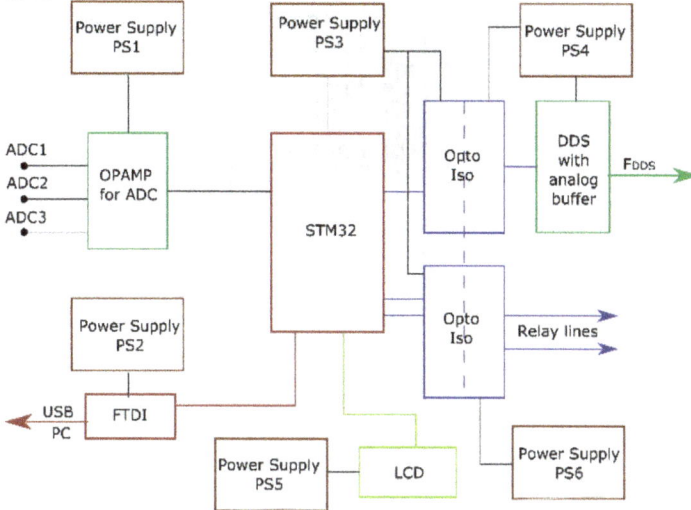

Figure 4. A block diagram of the developed system.

The main applications of the designed system are:
- service of piezoelectric or magnetic actuators/harvesters;
- scanning of the set frequency range using the actuator-harvester system with real-time performance reading;
- search and generation of the resonance frequency of mechanical structures;
- data transmission between the actuator and the harvester section in both directions;
- the ability to generate a signal for two actuators generating vibrations of the same frequency, but shifted in phase.

The results showing the frequency-amplitude characteristics for the circuit shown in Fig. 3 are shown in Fig. 5. It is worth noting that the highest voltage value (the highest efficiency) was obtained for the frequency in the above acoustic band (above 20 kHz). The zone marked SW in the figure defines the acceptable range of resonant frequencies and is about 20 kHz. The dashed line shows the voltage value at the level of 2.5 V, above which the microprocessor system was lost. In addition, point A indicate the most favorable frequency ranges for broadcasting information. As you can see, several frequency ranges can be distinguished that allow the system to be powered and depending on the conditions and needs, choices can be made between the required ranges of carrier signals. It should also be mentioned that it was possible to connect more microprocessors to the harvesters network, but in this case, it was necessary to maintain a strict order of their activation. This solution allows the system to be used in SHM (Structural Health Monitoring) applications with multiple sensors.

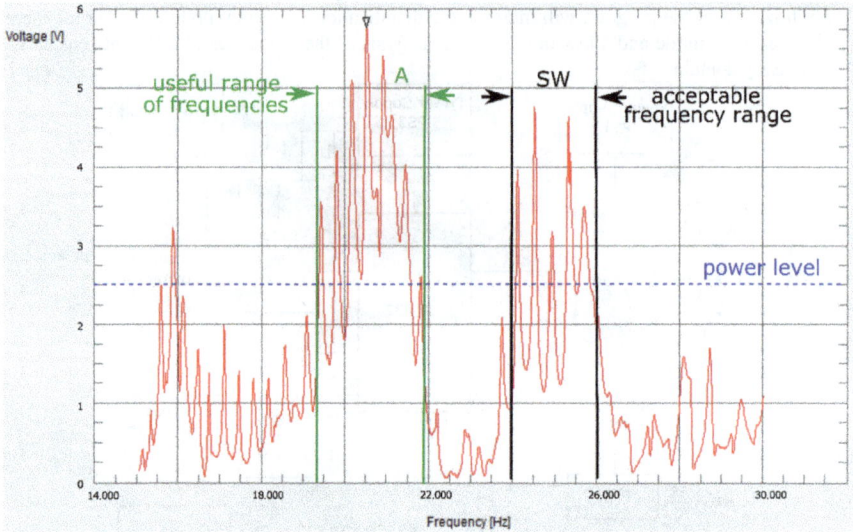

Figure 5. Frequency response of the dual-beam system.

The results obtained during the tests showed that the developed platform enables the transmission of energy over a distance of up to 3 m without the use of wires and using only various types of mechanical structures. This solution allows the use of various types of harvesters in many configurations, while the selection of the appropriate harvester system is influenced by the material and form of the medium through which energy and data are transmitted, as well as the ultrasound wavelength.

Conclusions

The article presents the results of work on the transfer of energy and information using various materials. The results achieved include:

- A platform that allows the transmission of power and information in relatively long rods using ultrasonic vibrations;
- The use of collectors/actuators based on both magnetostrictive and piezoelectric materials and the use of F2F (frequency/double frequency) procedures, which are a type of frequency modulation, to transfer information;
- Development of proprietary software for selecting the appropriate actuator and type of modulation as well as the recommended frequency band for energy and data transmission.

The results presented in the article are current and constitute the basis for further work in the field of energy and data transmission.

Acknowledgment

The research was founded within the pro-quality subsidy for the development of the research potential of the Faculty of Mechanical Engineering in 2022 "Excellence Initiative" - Research University (IDUB) 2022 of the Wrocław University of Science and Technology.

Materials Research Forum LLC
https://doi.org/10.21741/9781644902578-12

References

[1] A. Chandrakasan, R. Amirtharajah, J. Goodman, W. Rabiner, Trends in low power digital signal processing Circuits and Systems. In Proceedings of the 1998 International Symposium on Circuits and Systems (ISCAS), 4 (1998) 604-607.

[2] K. Mori, T. Horibe, S. Ishikawa, Y. Shindo, F. Narita, Characteristics of vibration energy harvesting using giant magnetostrictive cantilevers with resonant tuning, Smart Mater. Struct. 24 (2015) 125032. https://doi.org/10.1088/0964-1726/24/12/125032

[3] P. Loreti, A. Catini, M. De Luca, L. Bracciale, G. Gentile, C. Di Natale, The design of an energy harvesting wireless sensor node for tracking pink iguanas, Sensors 19 (2019) 985. https://doi.org/10.3390/s19050985

[4] F. Ait Aoudia, M. Gautier, M. Magno, O. Berder, L. Benini, Leveraging energy harvesting and wake-up receivers for long-term wireless sensor networks, Sensors 18 (2018) 1578. https://doi.org/10.3390/s18051578

[5] Y.C. Lai, Y.C. Hsiao, H.M. Wu, Z.L. Wang, Waterproof fabric-based multifunctional triboelectric nanogenerator for universally harvesting energy from raindrops, wind, and human motions and a self-powered sensors, Adv. Sci. 6 (2019) 1801883. https://doi.org/10.1002/advs.201801883

[6] S.X. Shi, Q.Q. Yue, Z.W. Zhang, A self-powered engine health monitoring system based on L-shaped wideband piezoelectric energy harvester. Micromachines 9 (2018) 629. https://doi.org/10.3390/mi9120629

[7] L. Wang, F.G. Yuan, Vibration energy harvesting by magnetostrictive material, Smart Mater. Struct. 17 (2008) 045009. https://doi.org/10.1088/0964-1726/17/4/045009

[8] S. Roundy, P.K. Wright, J. Rabaey, A study of low level vibrations as a power source for wireless sensor nodes, Comput. Commun. 26 (2003) 1131-1144. https://doi.org/10.1016/S0140-3664(02)00248-7

[9] S.R. Anton, H.A. Sodano, A review of power harvesting using piezoelectric materials (2003-2006), Smart Mater. Struct. 16 (2007), R1. https://doi.org/10.1088/0964-1726/16/3/R01

[10] P,D. Mitcheson, E.M. Yeatman, G.K. Rao, A.S. Holmes, T.C. Green, Energy harvesting from human and machine motion for wireless electronic devices, Proc. IEEE 96 (2008) 1457-1486. https://doi.org/10.1109/JPROC.2008.927494

[11] B. Viktor, Vibration energy harvesting using Galfenolbased transducer. In Proceedings of the SPIE Smart Structures and Materials + Nondestructive Evaluation and Health Monitoring, San Diego, CA, USA, 10-14 March 2013.

[12] Z. Deng, Nonlinear Modeling and Characterization of the Villari Effect and Model-Guided Development of Magnetostrictive Energy Harvesters and Dampers. Ph.D. Thesis, The Ohio State University, Columbus, OH, USA, 2015.

[13] Z. Deng, M. Dapino, Magnetic flux biasing of magnetostrictive sensors, Smart Mater. Struct. 26 (2017) 055027. https://doi.org/10.1088/1361-665X/aa688b

[14] H. Zhang, Power generation transducer from magnetostrictive materials, Appl. Phys. Lett. 98 (2011) 232505. https://doi.org/10.1063/1.3597222

[15] A. Viola, V. Franzitta, G. Cipriani, V. Dio, F.M. Raimondi, M. Trapanese, A magnetostrictive electric power generator for energy harvesting from traffic: Design and experimental verification, IEEE Trans. Magn. 51 (2015) 8208404. https://doi.org/10.1109/INTMAG.2015.7157483

Experimental Mechanics
Materials Research Proceedings 30 (2023) 83-90

Materials Research Forum LLC
https://doi.org/10.21741/9781644902578-12

[16] H. Liu, S. Wang, Y. Zhang, W. Wang, Study on the giant magnetostrictive vibration-power generation method for battery-less tire pressure monitoring system, Proc. Inst. Mech. Eng. Part C J. Mech. Eng. Sci. 229 (2014) 1639-1651. https://doi.org/10.1177/0954406214545821

[17] B. Yan, C. Zhang, L. Li, H. Zhang, S. Deng, Design and construction of magnetostrictive energy harvester for power generating floor systems, In Proceedings of the 2015 18th International Conference on Electrical Machines and Systems (ICEMS), Pattaya, Thailand, (2015) 409-412. https://doi.org/10.1109/ICEMS.2015.7385068

[18] B. Yan, C. Zhang, L. Li, Design and fabrication of a high-efficiency magnetostrictive energy harvester for highimpact vibration systems, IEEE Trans. Magn. 51 (2015) 8205404. https://doi.org/10.1109/TMAG.2015.2441295

[19] B. Nair, J.A. Nachlas, Z. Murphree, U.S. Patent US9438138B2. (2014).

[20] Y. Park, H. Kang, N.M. Wereley, Conceptual design of rotary magnetostrictive energy harvester, J. Appl. Phys. 115 (2014) 17E713. https://doi.org/10.1063/1.4865976

[21] M. Zucca, O. Bottauscio, C. Beatrice, A. Hadadian, F. Fiorillo, L. Martino, A study on energy harvesting by amorphous strips, IEEE Trans. Magn. 50 (2014) 8002104. https://doi.org/10.1109/TMAG.2014.2327169

[22] S. Kita, T. Ueno, S. Yamada, Improvement of force factor of magnetostrictive vibration power generator for high efficiency, J. Appl. Phys. 117 (2015) 17B508. https://doi.org/10.1063/1.4907237

[23] T. Ueno, Performance of improved magnetostrictive vibrational power generator, simple and high power output for practical applications, J. Appl. Phys. 117 (2015) 17A740. https://doi.org/10.1063/1.4917464

[24] Z. Deng, M.J. Dapino, Multiphysics modeling and design of Galfenol-based unimorph harvesters. In Proceedings of the SPIE Smart Structures and Materials + Nondestructive Evaluation and Health Monitoring, San Diego, CA, USA, (2015). https://doi.org/10.1117/12.2085550

[25] T.J. Lawry, K.R. Wilt, J.D. Ashdown, H.A. Scarton, G.J. Saulnier, A high-performance ultrasonic system for the simultaneous transmission of data and power through solid metal barriers, IEEE Trans. Ultrason. Ferroelectr. Freq. Control 60 (2012) 194-203. https://doi.org/10.1109/TUFFC.2013.2550

[26] S. Risquez, M. Woytasik, P. Coste, N. Isac, E. Lefeuvre, Additive fabrication of a 3D electrostatic energy harvesting microdevice designed to power a leadless pacemaker, Microsyst. Technol. 24 (2018) 5017-5026. https://doi.org/10.1007/s00542-018-3922-2

[27] Y. Liang, X. Zheng, Experimental researches on magneto-thermo-mechanical characterization of Terfenol-D, Acta Mech. Solida Sin. 20 (2007) 283-288. https://doi.org/10.1007/s10338-007-0733-x

[28] J. Kaleta, R. Mech, P. Wiewiórski, Development of Resonators with Reversible Magnetostrictive Effect for Applications as Actuators and Energy Harvesters; IntechOpen: London, UK, 2018. https://doi.org/10.5772/intechopen.78572

Experimental Mechanics
Materials Research Proceedings 30 (2023) 91-99

Materials Research Forum LLC
https://doi.org/10.21741/9781644902578-13

Canine hindlimb prosthetic research and its manufacturing with the help of additive technology

Michał Kowalik[1,a*], Malwina Ewa Kołodziejczak[2,b], Michał Staniszewski[1],
Mateusz Papis[1,c] and Witold Rządkowski[1,d]

[1]Institute of Aeronautics and Applied Mechanics, Faculty of Power and Aeronautical Engineering, Warsaw University of Technology, Nowowiejska 24, Warsaw, Poland

[2]Faculty of National Security, War Studies University, av. Antoniego Chruściela 103, Warsaw, Poland

[a]Michal.Kowalik@pw.edu.pl, [b]m.kolodziejczak@akademia.mil.pl, [c]Mateusz.Papis@pw.edu.pl, [d]Witold.Rzadkowski@pw.edu.pl

Keywords: Endoprosthesis, FEM Analysis, 3D Printing, Safety

Abstract. The purpose of this article is to develop a model and study an English bulldog endoprosthesis made with the help of additive techniques. An analysis of the literature shows the lack of such studies, including both additive techniques and topological optimization of prostheses. Before starting the actual study, general objectives were developed, which took into account the following: prosthesis assembly, fabrication technologies, fabrication conditions, shape, and range of work of the prosthesis. The dimensions of the prosthesis were identified based on the characteristics of the breed - the English bulldog. In the next step, the technology and construction materials were selected. The modeling of the prosthesis was based on the parameterization of dimensions. The parameters were linked by a skeletal model. Also, the objectives necessary to determine the factor of safety were defined. Boundary conditions were determined for the purpose of numerical calculations. The results in the form of reduced stresses and displacement distributions were presented on the maps. In the next part, topological optimization was performed, assuming the high stiffness of the system. Reduced stress maps and displacement distributions were generated for these results with the help of the FEM method. Validation of numerical calculations with real ones was performed.

Introduction

The Animal Protection Bill [1] as well as the EU directives point out that an animal is a living being capable of feeling pain and suffering. Man owes it respect, protection, and care. The animal should be treated humanely, that is, in such a way that ensures that its existential needs are met. In special cases, it is possible to kill the animal immediately. This possibility applies when an animal endures suffering and pain. Man must reduce the suffering of the animals and ensure their normal functioning [2]. One of the ways, in the case of dysfunctions connected with mobility, is the prosthetics. By applying modern manufacturing technologies [3], such as additive technology, it is possible to quickly and cheaply make individual prosthetics designed for a certain breed and even an individual animal [4]. It has also been found that when used for dogs, the endoprosthesis can prolong their lifespan [5].

The literature on the subject presents research on endoprostheses for dogs, made of various construction materials. One of the first endoprostheses design solutions was made of metal - a titanium-nickel prosthesis [6]. The simplest solutions present endoprosthesis as a metal monolith [7]. Other solutions include titanium endoprosthesis, made with a laser powder bed synthesis system, whereas, the cutting guides are made of ABS in a fusion deposition modeling system [8]. Not long ago, additive technologies started to be used to print this type of element (additive manufacturing) – 3D print [9], [10]. The literature fails to provide an additive manufacturing approach to animal prostheses that would apply topological optimization and their comparative studies. This paper presents it as a scientific innovation.

The purpose of this article is to develop a model and study an English bulldog endoprosthesis made with the help of additive techniques.

General objectives applicable to the conducted research

The model of the prosthesis will be limited by the research objectives. It was assumed that:

- the prosthesis is assembled to an endoprosthesis with a protruding stem ending in a thread or quick-connect;
- the prosthesis is made of plastic and printed by a 3D printer;
- the prosthesis is made in non-laboratory or industrial conditions;
- the material behaves isotropically;
- it should be easy to disassemble the limb replacement for maintenance purposes (e.g. cleaning);
- the shape of the limb is parameterized, that is, the shape is adaptable to the set overall dimensions of the bone (length of the femur, tibia, and metatarsus);
- the limb works on a single plane;
- the prosthesis is printed by a printer with a working scope of 220x220x230 mm;
- the prosthesis is waterproof;
- easily fixed;
- easily cleaned.

Due to the large breed diversity and the associated variation in the skeletal system of dogs, it is not possible to design a single prosthesis for all species [11], [12]. Therefore, for the purpose of this study, it was assumed that the prosthesis will be designed for the English bulldog. This dog breed is particularly prone to hip dysplasia, which requires endoprosthesis [13], [14]. The following dog characteristic was applied: weight 23-25 kg, height ca. 40 cm. As regards the shape of the hindlimbs, they are very muscular along their entire length, slightly shorter than the forelimbs. Knees are pointing slightly outward, ankle joints are small with a small angular range, and the paws are round and compact. Gait characteristics are as follows: short, quick steps, hind paws are not lifted high, on the contrary, they are low-lifted, almost gliding on the ground.

Prosthesis manufacturing technology and selected materials

The choice of prosthesis manufacturing technology depends on surface quality, mechanical strength, and material used. Because of the printed material, LOM (Laminated Object Manufacturing) technology has been ruled out [15], as the prosthesis cannot be made of paper, due to its poor strength properties as well as no water resistance, which then leads to a lack of durability. Among the technologies discussed, those that use other materials for printing purposes include SLS, FDM, and SLA. Analyzing the advantages and disadvantages of the aforementioned technologies, FDM technology was selected as preferable in this project because it is popular and available on the market, hence the lower costs in the case of mass production. The second argument behind choosing this method is the possibility to print large sizes, and thus produce monoliths.

Due to the high mechanical strength and increased fatigue strength, in order to meet the design objectives, PET-G material was selected as the printing material representing the properties summarized in Table 1.

Experimental Mechanics   Materials Research Forum LLC
Materials Research Proceedings 30 (2023) 91-99   https://doi.org/10.21741/9781644902578-13

Table 1. PET-G properties

Young modulus	2150 [MPa]
Tensile strength	50 [MPa]
Elongation at rupture	120 [%]
Density	1.27 [g/cm^3]

It is characterized by low hygroscopicity, it has been approved for contact with food, and for this reason, it has been used to manufacture dishes or bottles. As regards printing, it is characterized by a high level of adhesion during processing. This material behaves fine during mechanical processing and unlike PLA, PET-G printing is slower and more sensitive to parameter changes, which becomes immediately apparent in the printed detail, for example, in the form of lost transparency. The plastic is rigid and is used in prototyping products and verifying their mechanical strength. An additional advantage is the possibility to work in a wide temperature range (from -40ºC to +110ºC) and fine sliding properties, proven by a vast array of applications in sliding bearings. Glycol modification (suffix -G) increases Young's modulus.

To ensure the highest strength parameters, it was decided that the layers would be printed at 230ºC. A higher temperature (240ºC) should be applied to the first layer to ensure adhesion of the plastic to the heating table layer, a common problem that can lead to damaged prints. Printing the remaining layers at temperatures higher than 230ºC reduces the viscosity of the plastic as well as the dimensional accuracy of the printed part. As for printing speeds, the values of these parameters have been selected experimentally. In order to make sure that the layers of PET-G material fuse together, it is important to apply the correct speed. The speed cannot be too high or too low, as slow speed increases the time necessary to complete the print and can cause deformation caused by local overheating of the layer. The experimentally selected speeds are as follows: 30 mm/s for the outer walls and 50 mm/s for the inside. These parameters will help to maintain the consistency of the solid as well as the required dimensional accuracy.

Prosthesis modeling
The Skeleton model of the prosthesis was prepared in CAD Autodesk Inventor. First, it was necessary to prepare a simplified sketch, which included control dimensions and showed the kinematics of the device's operation, which will account for the limiting cases of the prosthesis. The geometric arrangement of the components on the plane helps to early predict the potential design problems such as where to set joint flexion limits or where to mount vibration dampers and other dynamic loads.

In order to make sure that the model functions properly, it was necessary to parameterize it. The next step involved setting limiting parameter values in Autodesk Inventor, from the smallest, allowing the model to be resolved, to the largest, which are optimal for the structure from a mechanical point of view. The parameters are linked directly to the skeleton model, which in turn translates these data into parameters for individual parts. Seven parameters were set relating to the length and three parameters relating to the angle.

Prosthesis' load
According to the chosen model, at rest the prosthesis is loaded evenly, i.e. each limb carries weight of ¼ of the dog's weight. This shows us that the forces and moments acting upon the prosthesis come from the weight and the response of the ground on which the dog is standing. According to the gait analysis, dynamic forces are acting upon the hindlimbs because the limb presses against the ground. Dynamic forces depend directly on the speed of the dog's movement, as well as the height to which the dog raises the limb. Depending on the speed of movement as well as the weight of the dog, the calculations will take into account an appropriate safety factor to compensate for the energy that is connected with the limb hitting the ground.

Experimental Mechanics   Materials Research Forum LLC
Materials Research Proceedings 30 (2023) 91-99   https://doi.org/10.21741/9781644902578-13

The following assumptions were made to determine the safety factor:
- approximation of the limb to a material point at the point of contact between the prosthesis and the ground;
- the ground does not absorb the impact energy;
- the mass of the material point is ¼ the mass of the dog m=6.25 kg;
- the maximum linear velocity of the limb from the initial to the final position is 2.78 m/s (about 10 km/h);
- the leg's elevation angle dφ=26°=0.4538 rad;
- the radius of the material point r=295.8 mm.

The analytically calculated coefficient is K=2.94.

Results of numerical calculations of the prosthesis

In order to complete numeric calculations, the following parameters were assumed:
- material is isotropic;
- main load F=179.92 N;
- secondary load F_{prot}=2.83 N, it is the force resulting from the mass of the prosthesis;
- the prosthesis is restrained at the contact surface of the prosthesis with the ground;
- the main load is applied to the surface where the endoprosthesis' mounting hole is located.

The finite element mesh was constructed from 10-node spatial (three-dimensional) elements. For static calculations, it was assumed that the mesh represented a characteristic size of 5 mm, with automatic compaction near notches. For models subject to shape analysis, the element size was reduced to 1 mm to obtain a higher resolution of the resulting shape.

During the process of designing the prosthesis, some of the dimensions were determined experimentally. The Shape Generator optimization module was used for further correction. The numerical calculations' results show us the maximum displacements of 1.848 mm and a minimum safety factor of 1.73. The actual load of the printed model points out to a low elastic deformability of the prosthesis, which is an undesirable phenomenon because of the transfer of loads coming from the ground's response to the endoprosthesis without their partial suppressing, which can in consequence cause damage to the prosthesis due to unpredictable rubbing at the junction or, in the worst case scenario, damage to the bone in the place where it is connected with the endoprosthesis.

The results of the calculations have been summarized in Table 2.

Table 2. Results of numerical calculations of the prosthesis

Reduced stresses at nominal load	31.4 [MPa]
Minimum safety factor	1.73
Maximum displacements	1.848 [mm]

The results of the calculations have been shown in the form of maps (Fig. 1).

Experimental Mechanics
Materials Research Proceedings 30 (2023) 91-99

Materials Research Forum LLC
https://doi.org/10.21741/9781644902578-13

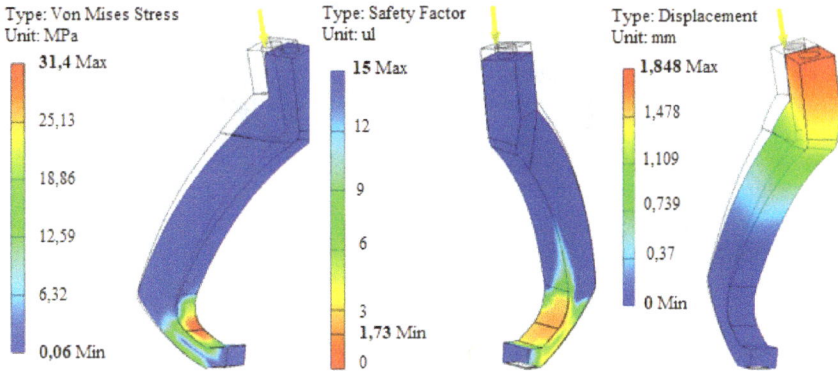

a) reduced stresses   b) distribution of safety factor   c) displacements

Fig. 1. Results of numerical calculations of the prosthesis

First, the optimization tool available in Autodesk Inventor was used, followed by numerical calculations, and in the end, the Shape Generator module was applied. The module has been designed to reduce the mass based on set criteria. Generating similar results as in strength calculations, it determines which finite element can be removed in order to slim down the structure to avoid damage.

For the purpose of optimization, an optimization area has been identified, defined as places of structural importance where it is not necessary to optimize the structure in order to slim it down. This step is important because the Shape Generator module is not able to recognize if the set part of the model is to fit and work with another detail or is normalized. In this project, this is the location where the prosthesis is connected and assembled with the endoprosthesis, as well as the place of contact of the prosthesis with the ground.

In this case, the criterion followed during the optimization in the module was stiffness maximization (Maximize Stiffness) and a weight reduction of 30-40%. A greater reduction is not necessary, since the weight of the prosthesis of ca. 200 g is almost imperceptible by the dog using it.

The precision of the operation was determined, as in every other process using the finite element method [16], [17]. Increased precision and accuracy increase the demand for computational resources and the time it takes for optimization to take place. The insufficient resolution will result in oversimplifications that can cause the structure to lose its stability, while a resolution that is too high increases computation time to several hours.

Optimization involves removing finite elements that do not carry loads or carry a relatively small value from the grid. Next, it reshapes it to distribute the stresses as evenly as possible. The result is a solid of an irregular shape caused by the removal of reduced finite elements. This function meets the condition of material continuity and the set conditions. The results of the optimization have been shown in Fig. 2.

Experimental Mechanics

Materials Research Proceedings 30 (2023) 91-99

Materials Research Forum LLC

https://doi.org/10.21741/9781644902578-13

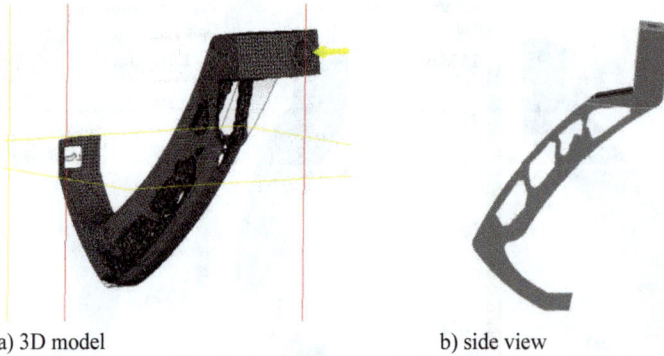

a) 3D model   b) side view

Fig. 2. A solid generated after prosthesis optimization

The results of numerical calculations after optimization have been summarized in Table 3.

Table 3. Results of numerical calculations of the prosthesis after optimization

Reduced stresses at nominal load	33.12 [MPa]
Minimum safety factor	1.64
Maximum displacements	2.415 [mm]

The results of the prosthesis calculation after optimization have been shown in Fig. 3.

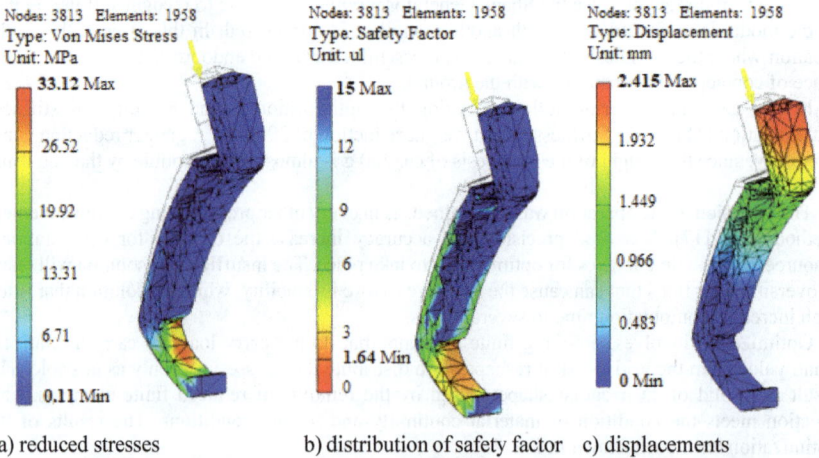

a) reduced stresses   b) distribution of safety factor   c) displacements

Fig. 3. Results of numerical calculations of the prosthesis after optimization

Experimental Mechanics   Materials Research Forum LLC
Materials Research Proceedings 30 (2023) 91-99   https://doi.org/10.21741/9781644902578-13

Empirical validation of numerical calculations of the prosthesis

To verify the numerical calculations, empirical validation of the calculations was performed. For this purpose, the printed prosthesis was mounted on an aluminum profile (Fig. 4). Followed by referential measurements. The prosthesis was loaded with cast iron weights of 17 kg (the weight of the tendon was omitted).

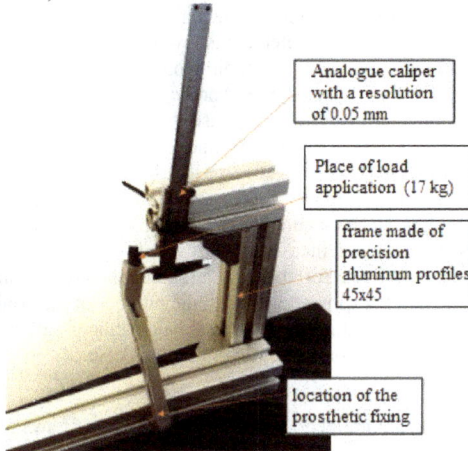

Fig. 4. Displacement measurements stand

Verifying the numerical calculations with the actual load applied, the relative errors of the maximum displacement measurement (against the numerical calculations) were calculated analytically, they were 94.8% for the prosthesis before the optimization process and 86.3% for the prosthesis whose shape was subject to the shape optimization process, respectively.

Conclusions

Based on the results of the conducted study, the following conclusions were drawn:

1. Excessive stiffness of the structure may cause the entire load to transfer to the endoprosthesis. Recurring loads with such stiffness cause the endoprosthesis and bones to carry increased loads from the ground, which may cause permanent damage to healthy parts of the dog's body.

2. Failure to install the endoprosthesis in the hole of the printed component correctly or improper matching of mating components may cause damage to the prosthesis because of the incorrect direction of loads. PET-G plastic is a relatively soft material and with poor metal-plastic mating, it will cause depravity of the latter.

3. The prosthesis is sensitive to the changing loads connected with the dog's weight. In this case, the dynamic load resulting from the dog's movement is three times greater than the static load. It means that each additional kilogram of the dog's weight will increase the basic dynamic load by about 30N.

4. To ensure optimal use of the material's properties, it is crucial to select the printer's plastic processing parameters. A printing temperature that is too low will prevent the layers from fusing properly and will eventually cause defusing, while a temperature that is too high will overheat the material or lead to incorrect application.

5. Optimization for the Maximize Stiffness criterion reduces maximum displacements and increases the safety factor through a more favorable distribution of loads within the solid. In addition to a more favorable distribution of stresses and displacements, optimization reduces the weight of the structure by 40%.

6. The relative measurement errors occur because the applied load differed from the nominal one by 8N. Secondly, it was assumed that the material would be isotropic. The third reason was that the strength of the printed element is lower than the strength of the same element made by injection molding, and, depending on the technology, is 60%-90% of the temporary endurance of the injected part. Another reason for the measurement error. The mounting method offered a different response than the assumed restraint of the contact surface of the detail with the substrate. Another reason for the error was a different load distribution along the layer lines, preventing a distribution that would match the numerical calculations. Additionally, another cause was the difference between Young's modulus used for the purpose of the calculations and the actual one. The sum of these errors makes up the total relative error of the measurement results.

7. In order to reduce the relative error, the material model should be changed to an anisotropic one, but in the case of FDM printed material, the number of material constants is so high that it cannot be implemented in the numerical calculation program that was used, so the assumed isotropy is a reasonable simplification in this case.

References

[1] Act of 21 August 1997 on the Protection of Animals (Journal of Laws of 1997, No. 111, item 724). In Polish

[2] M. Takayama-Ito, et al., Reduction of animal suffering in rabies vaccine potency testing by introduction of humane endpoints, Biologicals 46, pp. 38-45, 2017. https://doi.org/10.1016/j.biologicals.2016.12.007

[3] R. Bielawski, W. Rządkowski, M. P. Kowalik, M. Kłonica, Safety of aircraft structures in the context of composite element connection, Int. Rev. Aerosp. Eng. 13(5), pp. 159-164, 2020. https://doi.org/10.15866/irease.v13i5.18805

[4] R. Di Francia, V. Bizaoui, What if 3D Printing and Medicine had a dedicated Journal?, Ann. 3D Print. Med. 1, paper no 100007, 2021. https://doi.org/10.1016/j.stlm.2021.100007

[5] J. M. Liptak, W. S. Dernell, N. Ehrhart, M. H. Lafferty, G. J. Monteith, S. J. Withrow, Cortical Allograft and Endoprosthesis for Limb-Sparing Surgery in Dogs with Distal Radial Osteosarcoma: A Prospective Clinical Comparison of Two Different Limb-Sparing Techniques, Vet. Surg., 35 (6), pp. 518-533, 2006. https://doi.org/10.1111/j.1532-950X.2006.00185.x

[6] C. Sutton, et al., Titanium-nickel intravascular endoprosthesis: a 2-year study in dogs, Am. J. Roentgenol. 151 (3), pp. 597-601, 1988. https://doi.org/10.2214/ajr.151.3.597

[7] K. E. Mitchell, Metal endoprostheses for limb salvage surgery in dogs with distal radial osteosarcoma: evaluation of first and second generation metal endoprostheses and investigation of a novel endoprosthesis. 2017

[8] A. Timercan, V. Brailovski, Y. Petit, B. Lussier, B. Séguin, Personalized 3D-printed endoprostheses for limb sparing in dogs: Modeling and in vitro testing, Med. Eng. Phys. 71, pp. 17-29, 2019. https://doi.org/10.1016/j.medengphy.2019.07.005

[9] R. Mendaza-DeCal, S. Peso-Fernandez, J. Rodriguez-Quiros, Test of Designing and Manufacturing a Polyether Ether Ketone Endoprosthesis for Canine Extremities by 3D Printing, Front. Mech. Eng. 7, 2021. https://doi.org/10.3389/fmech.2021.693436

Experimental Mechanics
Materials Research Proceedings 30 (2023) 91-99

Materials Research Forum LLC
https://doi.org/10.21741/9781644902578-13

[10] S.-Y. Park et al., Custom-made artificial eyes using 3D printing for dogs: A preliminary study, PLoS One 15(11), paper no e0242274, 2020. https://doi.org/10.1371/journal.pone.0242274

[11] P. A. Manley, R. Vanderby, S. Kohles, M. D. Markel, J. P. Heiner, Alterations in femoral strain, micromotion, cortical geometry, cortical porosity, and bony ingrowth in uncemented collared and collarless prostheses in the dog, J. Arthroplasty, 10 (1), pp. 63-73, 1995. https://doi.org/10.1016/S0883-5403(05)80102-0

[12] E. Chisci, P. Dalla Caneva, i S. Michelagnoli, The "Dog Bone" Technique to Occlude a Branch Intentionally, Eur. J. Vasc. Endovasc. Surg. 61 (6), paper no 1035, 2021. https://doi.org/10.1016/j.ejvs.2021.02.032

[13] C. Ors, R. Caylak, E. Togrul, Total Hip Arthroplasty With the Wagner Cone Femoral Stem in Patients With Crowe IV Developmental Dysplasia of the Hip: A Retrospective Study, J. Arthroplasty 37 (1), pp. 103-109, 2022. https://doi.org/10.1016/j.arth.2021.09.007

[14] A. Santana, S. Alves-Pimenta, J. Martins, B. Colaço, M. Ginja, Imaging diagnosis of canine hip dysplasia with and without human exposure to ionizing radiation, Vet. J. 276, paper no 105745, 2021. https://doi.org/10.1016/j.tvjl.2021.105745

[15] G. Zhang, et. al., Frozen slurry-based laminated object manufacturing to fabricate porous ceramic with oriented lamellar structure, J. Eur. Ceram. Soc. 38 (11), pp. 4014-4019, wrz. 2018. https://doi.org/10.1016/j.jeurceramsoc.2018.04.032

[16] P. Różyło, H. Dębski, The Influence of Composite Lay-Up on the Stability of a Structure with Closed Section, Adv. Sci. Technol. Res. J. 16 (1), pp. 260-265, 2022. https://doi.org/10.12913/22998624/145156

[17] Z. Pater, P. Walczuk-Gągała, Conception of Hollow Axles Forming by Skew Rolling with Moving Mandrel, Adv. Sci. Technol. Res. J. 15(3), pp. 146-154, 2021. https://doi.org/10.12913/22998624/139134

Keyword Index

About the Editors

Professor Paweł Pyrzanowski, employed at Warsaw University of Technology, Faculty of Power and Aeronautical Engineering.

Member of many associations related to experimental mechanics:

- Polish Association of Experimental Mechanics – Founding member, Chairman of the Board since 2018;
- The Mechanics Committee of the Polish Academy of Sciences – Member, Chairman of the Experimental Mechanics Section – since 2016;
- Danubia-Adria Society for Experimental Mechanics – Member - representative of Poland since 2016;
- Symposium on Experimental Mechanics in memory of prof. Jacek Stupnicki – Chairman since 2012.

Nominated as a professor in 2020. Author of more than 90 publications (monographs and chapters, scientific magazines and conference papers).

Mateusz Papis, PhD, employed at Warsaw University of Technology, Faculty of Power and Aeronautical Engineering.

Member of Symposium on Experimental Mechanics Organizing Committee since 2018.

PhD degree from 2021. His scientific interests include reliability and safety of systems, risk analysis. Author of more than 10 publications in these fields of study.

www.ingramcontent.com/pod-product-compliance
Lightning Source LLC
Chambersburg PA
CBHW071719210326
41597CB00017B/2531